THE
INNOVATOR'S
HYPOTHESIS

THE INNOVATOR'S HYPOTHESIS

HOW CHEAP EXPERIMENTS ARE WORTH MORE THAN GOOD IDEAS

MICHAEL SCHRAGE

THE MIT PRESS   CAMBRIDGE, MASSACHUSETTS   LONDON, ENGLAND

MIT Press books may be purchased at special quantity discounts for business or sales promotional use. For information, please email special_sales@mitpress.mit.edu

This book was set in Sabon and Futura by the MIT Press. Printed and bound in the United States of America.

Library of Congress Cataloging-in-Publication Data

Schrage, Michael.
The innovator's hypothesis : how cheap experiments are worth more than good ideas / Michael Schrage.
    pages cm
Includes bibliographical references and index.
ISBN 978-0-262-02836-3 (hardcover : alk. paper) 1. Technological innovations.
2. Creative ability in business. I. Title.
HD45.S367 2014
658.4'063—dc23
                    2014013729

10   9   8   7   6   5   4   3   2   1

CONTENTS

PREFACE

This is a modest book with an immodest argument: creative experimentation, with and within constraints, makes high-impact innovation a safer, smarter, simpler and more successful investment. It didn't start out that way; I had to experiment.

Back in 1999, the Harvard Business School Press published *Serious Play: How the World's Best Companies Simulate to Innovate*, my book exploring how models, prototypes, and simulations shape innovation culture. The book proved unexpectedly influential. Companies all over the world invited me to work with them. I got unusually serious follow-ups from executive education classes and workshops at MIT. The opportunities were remarkable, the impact terrific.

But I soon discovered something weird. Although they loved using models, prototypes, and simulations to "innovate out loud," most organizations seemed less interested in detailed "deep dives." Yes, technology companies wanted more collaborative prototyping, and professional service firms sought more dynamic visual simulations. But the majority wanted something different. They wanted their people innovating much faster, much better, and much cheaper. They wanted the organization's best minds and brightest talents engaged in new ways. Enhancing human capital mattered as much as creating new products, services, or both. They desired effective change, not disruptive revolution. These companies craved a simple, fast, and frugal innovation capability.

I confess: my immediate reaction was to repurpose my material and expertise. I'd better tailor and customize "rapid prototyping" and "rapid simulation" workshops to their briefs. But that frankly wasn't good enough. While modeling and prototyping were great for engineers, developers, and more technical folk, they lacked broader enterprise appeal. I listened more carefully to people's innovation ambitions, fears, and constraints. I realized I had to reinvent, not just repurpose.

What could businesspeople collaboratively design to create or discover new innovation insights? Experiments. Business experiments. Get small teams from across the enterprise crafting business experiments that make their top managements sit up and lean forward. Push small teams to work collectively to come up with business hypotheses inspiring urgency and curiosity.

But this wouldn't be "blue sky/anything goes" experimentation. This would be experimentation within constraints: No blank checks, no unlimited budgets. No "innovation vacations" for meditation and contemplation. Friendly rivalry between small teams. Clear deadlines. Explicit deliverables to top management. Opportunities to "experiment out loud" in ways commanding strategic attention and respect. These were fresh ingredients for cultural change and market impact. These improvised constraints evolved into the 5×5 framework presented in this book.

My initial clients and classes couldn't have been more open or receptive. Talented, ambitious people enjoy demonstrating how collaborative and innovative they can be. One of the finest compliments I ever received came from an initial 5×5 skeptic: "When my team first started designing experiments, the exercise seemed ridiculous to me. Now it seems obvious."

Of course, there were disappointments and outright failures. Some teams never grasp the meaning of experimentation or get beyond design clichés (i.e., cutting prices, advertising more, etc.). But the methodology's fast and frugal approach keeps the costs of failure low and the opportunities for organizational learning high.

The ongoing success of 5×5 in organizations worldwide reflects two fundamental shifts that have transformed my teaching, consulting, and advisory work since *Serious Play*. The first is transitioning from the practice of selling solutions to the promise of providing an effective approach. The second is moving from the "transmission of expertise" toward the "cultivation of capability."

Those sensibilities have challenged me to pay closer attention to how people learn, rather than what I should teach. *The Innovators' Hypothesis* is rooted in observing what makes 5×5 teams effective.

Providing "the answer" is not the answer. Too many "solutions" either don't solve the problem or don't solve the problem for long. They are frequently too brittle, too complicated, or too customized to cost-effectively adapt to changing circumstance. I prefer presenting clients and students with "approaches"—that is, methods, tools, and frameworks that put greater power in their hands and minds. That's more sustainable.

So I help organizations design fast, frugal, and high-impact business experiments that make innovation simpler, safer, and scalable. That's my mission statement. That's the goal and purpose of this book. The operational definitions of *fast, frugal,* and *high-impact* constantly change. So do the techniques and technologies of simple, safe, and scalable. There's arguably never been a better time in history to explore and exploit creative business experiments.

When I co-organized an MIT workshop on exponential experimentation in 2009, it was blindingly obvious to everyone in that room that experimentation's true economic potential had barely been tapped. I was so excited that I started writing a book—this book—on the subject. It was going to be a Grand Tour of the experimental future. I started off very strong—but then I stopped. I had a serious crisis of confidence. I doubted myself and what I was doing. I felt I was giving a really good answer to exactly the wrong question. Faster, better, cheaper, and simpler experiments aren't the goal; they're means to an end.

But what should that end be? That question haunted me. The answer is clearer for physics, chemistry, and biology than for industries, companies, and businesses. The more 5×5's I facilitated, the more organizations I helped run faster, better, and cheaper experiments, the more uncomfortable I became. Yes, we were coming up with terrific hypotheses and experiments. Yes, we were transforming the innovation conversation. But I felt consumed more by the tactics and techniques of experimentation than what experiments should ultimately be for. So I stopped writing. I quit this book and started another. That book made this book possible.

Who Do You Want Your Customers to Become? was published by Harvard Business Review Press in 2012. Its essential insight is that innovation is an investment in the human capital, capabilities, and competencies of customers and clients.

Business history gives great credence to this "human capital" model of innovation. For example, George Eastman didn't just invent cheap cameras and film; he created photographers. Steve Jobs didn't merely "reinvent" personal computing and mobile telephony; he reinvented how people physically touched and talked with their technologies. Successful innovators have a "vision of the customer future" that matters every bit as much as their product or service vision. By treating innovation as an investment in customer futures, organizations can make their customers more valuable. In other words, "Making Customers Better Makes Better Customers."

This was the breakthrough I needed to finish this modest book with its immodest argument. Experimentation didn't just offer the best way to cost-effectively invest in innovation, it offered the most innovative way of exploring how to invest in the human capital, competences and capabilities of one's customers. Organizations should do faster, cheaper, and simpler experiments to make smarter, better, and safer innovation investments in their customers' futures. If you know "who you want your customers to become," there's no better methodology than the 5×5 for identifying high-impact innovation investments. If you don't know, then the 5×5 is ideal for the experimentation and discovery you need to find out. The 5×5 is ideal for empowering organizations to experiment with experimentation.

What I've rediscovered is that *The Innovator's Hypothesis* is a book about transforming the human capital and capabilities of both you and your customers. The more simply and creatively you can experiment, the more simply and creatively you can learn to add value to your customers. The more simply and creatively your customers can experiment with you, the more valuable you can become to them. As a 5×5 team member once remarked, "The bottom line is a learning curve." Yes it is. And you should use this book to experiment with it.

ACKNOWLEDGMENTS

Acknowledgments could just as accurately be called "gratitudes." The development of the 5×5 and this book wouldn't have been possible without the enthusiasm, skepticism, creativity, and curiosity of my students and clients. I'm enormously grateful to them. The cliché that the best way to learn a subject is to teach it could not be truer. We learned together. We shared frustration and disappointment and mistakes. We also got results that mattered. Their energy and ingenuity was humbling, impressive, and exciting.

That energy and ingenuity was also global: we ran scores of 5×5 X-team workshops, classes, and engagements worldwide: in Russia, China, India, Australia, Brazil, Colombia, Chile, Japan, the UK, Europe, and, of course, North America. The world is filled with authentically smart people looking to better innovate, collaborate, and experiment with each other. That's both exciting and inviting. I can't overstate either my pleasure or relief at how well students and clients alike committed themselves to fast, frugal, and simple experiments with fast, frugal, and simple experimentation. They made the methodology real; they made it better; and they made it their own. The best ideas and insights in this book aren't mine; they're ours.

MIT was this book's institutional incubator. What's more *mens et manus* than an experiment? I'm particularly grateful to Erik Brynjolfsson and David Verrill of the Sloan School's Center for Digital Business. They've been generous and supportive colleagues. Erik and I cowrote an article of digital experiments and the Center sponsored what turned out to be an influential workshop on "business experimentation" in 2009. The CDB's Tammy Buzzell's coordinative prowess was essential to its success.

The opportunity to iteratively redefine and redesign the 5×5 through the Sloan School's Executive Education programs also proved indispensable. I owe thanks to Peter Hirst, Tommy Long, and Court Chilton. The MIT Professional Education summer sessions on innovation taught with Harvey Sapolsky and Sandy Wiener over the past 15 years were similarly stimulating and helpful. I appreciate Bill Aulet, director of the Martin Trust Center for MIT Entrepreneurship, inviting me to bring the methodology to his entrepreneurs.

MIT's Industrial Liaison Program proved a remarkable resource for connecting worldwide to organizations open to innovating with experimentation. Admittedly, global organizations coming to MIT for "innovation inspiration" and advice represent a biased sample. But working with innovators striving to become—or remain—"world class" raises everyone's game. Tony Knopp, Randall Wright, Todd Glickman, Marie van der Sande, Nenna Buck, Rachel Oberai-Soltz, Ken Goldman, Graham Rong, James Gado, Klaus Schleicher, Enrique Shadah, Irwin Winkler, and director Karl Koster have all facilitated introductions and interactions that, over the years, have profoundly influenced my experimental designs.

The Moscow School of Management Skolkovo became an unexpected laboratory for 5×5 research with its Russian-flavored cross-cultural mix of executive education programs, entrepreneurs,

MBAs and state-owned enterprises. My special thanks to Pavel Lushka, Svetlana Pashkevich, Olga Kuzmicheva, Violetta Grigorieva, Evgenia Feoktistova, and the Skolkovo community for their support.

Both directly and indirectly, this book is better because of conversations and/or arguments with Michael Cramer, APT's Jim Manzi, Tom Lockwood, Shipra Gill, BT's Steve Whittaker, Jennifer Newton, Melinda Merino, Rob Fulop, Qi Lu, Ronny Kohavi, Gary Loveman, Bob Buderi, Jason Pontin, Dan Ariely, Anne Milley, Richard Narramore (to whom I owe an apology), Steve Postrel, Reihan Salam, Louise Valentine, Andrew May, Andrew Marshall, Nicholas Negroponte, Tom Malone, Greg Linden, Eva Lee, and my brother Elliot. I apologize unreservedly to anyone and everyone I may have overlooked. In truth, there's no way I can adequately acknowledge the macro/micro contributions made both by accident and design to this work. What's equally true, however, is that any and all mistakes, errors and sins of omission/commission are ultimately my responsibility. I accept it with relief.

I've known Gita Manaktala, the editorial director of the MIT Press, longer than either of us might like to acknowledge. I've always known her as a friend, not a colleague. It's enormously gratifying (but not surprising) that she's wonderful as both. My editor, John Covell, has been patient, diligent, supportive, and scrupulously concerned with assuring that *The Innovator's Hypothesis* captures the best parts of my voice and insight. More painfully, he's been committed to imposing a structure that made this book narratively coherent, as well as useful. I hope he has not suffered too much. Similarly, I hope Marcy Ross, my production editor, has not found my obsessive compulsiveness or compulsive obsessiveness too daunting or draining. The entire MIT Press staff has stepped up to make *The Innovator's Hypothesis* one of its most interesting publishing experiments. I look forward to being even more grateful as the book goes through future editions.

But my most heartfelt gratitude and acknowledgments belong to my wife, Beth Ann. As a real-world CEO, she understands economic constraints, accountability, and talent. While I couldn't have written this book without my students, clients, and colleagues, I couldn't have enjoyed doing it as much without her caring, compassion, and—this is not a typo—impatience. She has my love and admiration—occasionally in that order.

THE
INNOVATOR'S
HYPOTHESIS

1 THE INNOVATOR'S VISION

The Innovator's Hypothesis champions simple, fast, and frugal experimentation as the smartest investment that serious innovators can make. The book describes a lightweight, high-impact methodology that clarifies and sharpens innovation focus, facilitates alignment between top-down strategic visions and bottom-up innovation empowerment, and encourages an "actions speak louder than words" innovation culture. This methodology works.

Both guide and manifesto, this book explains why good ideas are typically bad investments. A new economics of experimentation is transforming innovation investment opportunities worldwide. Simple, fast, and frugal experiments inspire creativity and capability in a way that better analytics and planning can't. In networked industries and global markets, experimentation has become as much an asset class as a core competence. It must be managed accordingly. So *The Innovator's Hypothesis* challenges innovators to become Warren Buffetts—fundamental value investors—of experimentation. What experiments buy a dollar's worth of inno-vation insight for 50 cents or 20 cents … or even less?

Drawing extensively from real-world examples, this book con-cludes that portfolios of simple experiments generate more valuable insights and innovation than volumes of sophisticated analyses. Little experiments make "big data" more useful and usable. Vol-taire once observed that "God is not on the side of the biggest battalions, but of the best shots." *The Innovator's Hypothesis* is about creating cultures of experimentation where innovators can take their best shots to succeed.

This book is for people seeking dramatically greater returns from their organizations' innovation investments. It is for leaders striving to radically improve, accelerate, and simplify how their people innovate; for organizations that need to manage innovation risk better; and for managers open to the reality that quick and simple experiments can be more valuable over a much shorter time frame than complex and comprehensive analyses.

In other words, this book is for the organizationally frustrat-ed—people who have to do much more with much less. Anyone who "owns"—or is accountable for—innovation initiatives will find fresh value here. *The Innovator's Hypothesis* is for people who know they need faster, better, and simpler tools to revisit and rethink the business fundamentals that matter most.

CEOs desire deeper insight into top talent ingenuity. Execu-tives want to be confident that their associates are more creatively capable than the organization permits. Leaders are exhausted by costly and sluggishly complicated innovation processes. CFOs are desperate for cleverer ways to assess and manage innovation risk. Intrapreneurs can't figure out how to test their best ideas quickly or cheaply. Entrepreneurs fervently fear that "only the paranoid survive." And iPad and Android addicts look at the combinatorial explosion of digital platforms and device and wonder, "Wow! Why aren't we doing more with this?!"

This book works for all these people, addressing their frustra-tions and opportunities explicitly. Its 5×5 methodology is ideal for innovators intent on making an impact bigger than their budgets. Constructively influencing innovation culture is hard. This book is written for change agents who sense that experimenting with experimenting can positively transform innovation culture, and shows them how to do just that.

THE 5×5

The 5×5 X-team approach is a rapid innovation methodology emphasizing lightweight, high-impact experimentation, as follows: Give a diverse team of 5 people no more than 5 days to come up with a portfolio of 5 business experiments that cost no more than $5,000 (each) and take no longer than 5 weeks to run. The willingness to ask simple questions is essential. Simplicity invites ingenuity. The 5×5 offers a fast, cheap, and ingenious method for innovators to revisit—and test—business fundamentals safely. Simple questions about customer segmentation, sales, pricing, performance, and language inspire successful, high-impact hypotheses.

Why this book? Innovation too often is too slow, too expensive, too complicated, too risky, too rigid, too dull, too little, and too late. With innovation, the status quo isn't good enough and isn't improving fast enough. Organizations overinvest in the wrong things and underinvest where they might get the greatest impact fastest. They need to be smarter and better innovation investors. But they need to invest smarter and better in faster, more frugal, and simpler ways.

The Innovator's Hypothesis identifies underappreciated and undervalued investment opportunities. These opportunities challenge organizations to rethink what kind of innovation investors they want to be. They encourage innovators to revisit how they want to create new value. Experiments can be assets. Experiment portfolios can become investment platforms where organizations quickly and cheaply explore real-world trade-offs between innovation risk and innovation reward.

By design, the 5×5 methodology renders the traditional metrics around costs and time irrelevant. Its focus on simplicity and speed effectively minimizes legacy concerns about scheduling and finance. The goal is to align people around the innovation insights generated by the experiments portfolios.

Think of the 5×5 framework as an experimental extension of Vilfredo Pareto's (1848–1923) famous 80/20 principle—the economic insight that much of the time, 80 percent of the outputs can be explained by roughly 20 percent of the inputs. (That is, 20 percent of the employees do 80 percent of the work, or 20 percent of the customers generate 80 percent of the revenue.) But where Pareto describes an 80/20 distribution, the 5×5 seeks an 80/20/20 vision.

That is, what hypothesis could we test—what experiments could we run—that generate 80 percent of the useful information that we need to make a decision in 20 percent of the time, and with but 20 percent of the resources that we ordinarily use to do so?

The 80/20/20 vision defines the conceptual constraints for 5×5 experimental designs. The goal isn't getting the best insight or the right answer or "95 percent of what we need to know"; it's designing and doing experiments that get us more than three-quarters of the way to what we think we need to understand in a fraction of the time and at a fifth of the cost of what we ordinarily invest to get that essential information. The 80/20/20 vision describes the portfolio parameters for superior returns on 5×5 investments.

There are many excellent books about innovation and innovation culture. There are many excellent books about experiments and experimentation. But what sets *The Innovator's Hypothesis* apart is its investment emphasis. The 5×5's fusion to the 80/20/20 vision describes not just an innovation methodology but a fundamental value investment philosophy—buy valuable insights cheap. Creating an innovation culture that respects and embraces a new economics of experimentation is a Big Deal.

Experimentation for the sake of experimentation is inefficient indulgence—not unlike managing for the sake of managing or cost-cutting for the sake of cutting costs. Experimentation should be a means to an end. Experiment portfolios should be connected to actionable insights and essentials. This book exists because the real innovation investment potential of experimentation has yet to be tapped.

This book is not a call to innovation revolution. It's not a leisurely meditation, exegesis, or disquisition on the history or epistemology of experimentation. This book doesn't discuss Ronald Fisher's concept of "design of experiments," factorial design, or Latin hypercube sampling. This isn't the book for detailed technical discussions or descriptions about what makes experiments work.

Why not? Because my work with organizations strongly suggests that revolutionaries are typically better at bridge burning and martyrdom than value creation. Most organizations lack the time and bandwidth to appreciate—let alone study—the historical context of experimental design in science or business. And, frankly, the overwhelming majority of general managers lack the quantitative or statistical numeracy to exploit effectively the computational tools that distinguish an Amazon, a Netflix, a Google, or a Capital One. At these businesses, experimentation is a cultural norm driven and demanded by their founders. That's who they are. Indeed, the cultural aberration and outlier there is *not* experimenting.

The organizations I've advised—the executive education workshops I've run—aren't hungry for revolution. To the contrary, they're looking for frameworks and tools that measurably add value before they threaten management. They're careful, not cautious. They want their commitment to cultural change to extend collaborative capabilities, not provoke new conflicts. They want people trained to do more with less without spending much more or taking too much time.

Similarly, they've learned the hard way that technical experimentation is not the same as business experimentation. They've discovered that statistical validity in experimental design is not the same as testing simple hypotheses to find and explore new value. As Ernest Rutherford, the great Nobel Prize–winning experimental physicist, remarked, "If your experiment requires statistics, you ought to have done a better experiment." This book emphasizes doing better experiments.

Yes, technical expertise is increasingly important to innovative experimentation. The future of experimenting with experimentation for the purpose of creating new value will demand superior quantitative knowledge and skills. But that knowledge and those skills will be underappreciated and underused without first following the tenet of *The Innovator's Hypothesis.*

This book's mission is to convince organizations to experiment with experimenting through the 5×5 methodology. Its goal is to get every 5×5 team to grasp the importance of the 80/20/20 vision when designing simple experiment portfolios. The hope is that organizations will quickly discover how fundamental experiment portfolio management creates more agile, more creative, and more successful innovation cultures—for less.

2 WHAT IS A BUSINESS HYPOTHESIS? WHAT ARE BUSINESS EXPERIMENTS?

THE BUSINESS HYPOTHESIS

For the purposes of this book and the 5×5 methodology, a *business hypothesis* is a testable belief about future value creation. It is not a search for truth or fundamental understanding; a business hypothesis suggests a possible and plausible causal relationship between a proposed action and an economically desirable outcome.

If there is not an explicit and understood measure for new value creation, it is not a business hypothesis. There must be a metric— revenue, margin, usage, satisfaction, engagement, etc.—for assessing value. What value metric does the business hypothesis seek to improve?

Structurally, a 5×5 business hypothesis takes the form of:

The Team Believes Exploring This <Action/This Capability>

Will Likely Result in This <Desirable Improvement/Outcome>

We'll Know This Because <Our Explicit/Understood Metric> <Significantly Changed>

If the business hypothesis isn't written down, agreed upon and readily shareable, it's not a business hypothesis.

A 5×5 business hypothesis does not live in the mind. If its essential argument isn't tweetable, it's likely not thought through with rigor and clarity. A crisply articulated business hypothesis makes simple, frugal and scalable business experiments easier to design and test.

THE BUSINESS EXPERIMENT

For the purposes of this book and its 5×5 methodology, a *business experiment is* an easily replicable test of a business hypothesis that generates meaningful learning and measurable outcomes. A business experiment neither validates nor proves a business hypothesis; it does not solve a business problem. It meaningfully and measurably provides some insight into the relationship—if any—between action and outcome. The importance and significance of that insight depends on the design, implementation, and interpretation of that simple experiment.

Easily replicable means that almost any group of individuals in the organization similar in background and experience to a 5×5 team could quickly grasp and confidently replicate the proposed experiment with minimal training or difficulty. There's no magic, nuance or subtlety required. The business experiment's purpose, design, and implementation are invitingly clear. *Meaningful learning* indicates that the 5×5 teams—and their colleagues—both individually and collectively agree that they acquired greater insight and awareness into the "value creation" relationship they sought to explore. That is, they gained valuable information about creating valuable information.

Measurable outcomes is the commitment to the agreed-upon "value metric" for assessing experimental effectiveness. Business experiments aren't about seeing if something "works," they're about successfully aligning business hypotheses with respected metrics.

The media of 5×5 experimentation don't matter—Facebook, websites, Post-it notes, QR codes, paper prototypes, Lego, or

FORMULATING THE HYPOTHESIS

One retailer identified "Leveraging Customer Touchpoints" as its design theme for business hypothesis and experiment. The organization sought to explore how to take advantage of—or add value to—every known point at which the store interacted with the customer. The broader hypothesis was that the store could improve customer experience, learn more, and possibly increase sales by "offering something extra" at key touchpoints.

A team identified "receipts" as a particularly important customer touchpoint. The team was interested in "digital/online receipts" that customers could elect to have sent to them after a purchase. The team argued that "receipts" shouldn't be seen as the confirmation/conclusion of a transaction but also as an invitation to continue interacting with the store. In other words, digital receipts could become "promotional platforms." How might that be accomplished?

The team formulated roughly the following hypothesis: "The team believes exploring an embedded hyperlink promising a coupon/discount/promotion/benefit into a digital receipt will likely result in customer 'click-through' to obtain the offers. We'll know this because we'll be able to measure the click-throughs."

Of course, the team discussed what kind of links to embed (for example, visual, textual, prominence, and color) and what kinds of offers to test (such as a discount on next purchase, email coupon, or special promotion). That quickly evolved into specific experiments. But there was swift consensus that the retailer should rethink and redesign "ereceipts" as touchpoints.

apps—are all fair game. The most important concern is how usefully the experimental design and media facilitate testing the business hypothesis:

- Does the experimental design elicit and illuminate the core value relationship the hypothesis proposes?
- Do the chosen experimental media enhance or constrict the possible learning?
- How compatible are the experimental media with the metrics?

For example, would both a digital and physical business experiment dramatically improve the learning and measurability of a high-impact hypotheses?

These questions demand business judgment, not just technical craft; these questions invite collaborative interaction, not just expert advice. Translating hypotheses into experiments; revisiting and revising the metrics that matter most; commitment to rapid iterative learning—these are behavioral shifts and important investments in facilitating an innovation culture in which actions speak louder than words.

MOVING FROM HYPOTHESIS TO EXPERIMENT

A provider of chemical engineering equipment identified "improving the efficiency and effectiveness of our onsite maintenance efforts" as its innovation theme for hypothesis and experiment. The firm calculated that onsite maintenance/replacement/repair was one of its most significant and time-consuming operational expenses. Misdiagnoses were not uncommon and, frequently, service people didn't always have the right equipment on hand to make needed repairs. These issues led to ripple effects and cascading delays. The resulting inefficiencies were not only expensive but also angered customers. This undermined credibility and reduced opportunities to "upsell" preventive maintenance and other services. The organization sought to dramatically reduce diagnostic error and time on customer site.

The business hypothesis centered on having service people learn more about the problem before they went to the site. The team formulated roughly the following hypothesis: "If customers can quickly and easily get us more information about the problem, we'll be able to bring the right equipment and know what we're fixing. What onsite information could the team see in advance that would make a difference?"

The team wondered whether customers could or would take mobile phone photos of the relevant site points and problems (pipes, dials, leaks, etc.) when the maintenance problems were called or emailed in.

The team believes exploring how to get customers to send relevant photos will likely result in a more detailed technical understanding of the maintenance problem before a visit. The company will know this because its service people will spend less time onsite and make dramatically fewer "additional equipment" requests.

This hypothesis led to further discussion—led by the services/maintenance group—on what photos and images would be most valuable. The team also discussed what "apps" might be downloaded to customer phones or tablets that would improve visual and/or acoustic diagnostics. The concept of using customers and their mobile devices as part of a "pre-visit technical service team" took hold.

But the simple experimentation began with learning if customers could and would email specifically requested photos of their onsite problems.

THE
INNOVATOR'S
FOCUS

3 IDEAS ARE THE ENEMY

Good ideas are usually bad investments. They're rarely worth the time, money, or effort. They overpromise and underdeliver. They seduce and they cheat. Good ideas are appealing because, well, who doesn't like good ideas?

By definition, good ideas are good; better ideas are even better; and great ideas are just great. Economists left, right, and center agree that successful businesses need "good ideas" to grow. Management icons from Peter Drucker to Jack Welch observe that great leaders must have the wisdom and courage to recognize and reward great ideas.

Reasonable and rational? That's the conventional wisdom. But in reality, it's harmful nonsense. Good ideas are actually the enemy of productivity. A focus on good ideas inflicts terrible damage—operational and emotional—on good managers and good businesses alike. Business ideaholics, not unlike meth addicts and other junkies, are always looking for the next fix. They crave the buzz, rush, or high that supposedly comes from injecting a really good idea into their managerial mainstream. Good ideas might be better described as the empty calories of enterprise innovation: accessible, tasty, and momentarily satisfying. But they're not good for you. They'll make you sick.

The idea that good ideas mostly aren't seems counterintuitive, but the evidence says otherwise. When organizations actually review economic returns on so-called good ideas, they typically discover that the accumulated costs seldom outweigh the anticipated benefits. Nobel Prize–winning economics research affirms that experts are particularly prone to overconfident predictions of success.

Even the most innovative firms find a disconcerting percentage of their cherished pet ideas actually annihilated large chunks of value. DuPont, Xerox, General Electric (GE), Sony, Citigroup, and General Motors (GM) are just a few companies where rigorous retrospective analyses affirm the high price—and higher costs—of good ideas gone bad. Multibillionaire superinvestor Warren Buffett—a businessman attuned to the potential value of ideas—has commented at great length on this unhappy managerial pathology.

The story gets worse. "Ugly duckling" ideas that top management dismisses as marginal may turn out to be beautifully profitable swans. Were these mediocrities that just got lucky, or were they ideas that, for whatever reason, management simply misanalyzed and misunderstood?

Organizations striving to innovate consistently find themselves unpleasantly pincered by probability: the value of good ideas is consistently overstated, while the possible value of potential winners is underestimated. Business history indicates that, on average, most organizations in most industries do an unimpressive job realizing respectable returns from good ideas. This is as true for software and silicon as it is for automobiles and airplanes. The economics of good ideas can be ugly.

Of course, great ideas—even good ideas—exist that positively transform enterprises. But honesty compels serious managers and investors to ask if those successes represent the rule or the exception. If producing good outcomes from good ideas isn't the investment norm, then just how good are those ideas? How should innovation-hungry enterprises value assets that, more often than not, fail to generate decent returns? Perhaps these alleged assets are better described as liabilities.

There is not a business executive alive who hasn't heard or uttered the phrase "Well, it seemed like a good idea at the time" in the wake of costly failures or pathetic results. Of course it did. It was supposed to. That's why good ideas are so reliably dangerous. They look like good investments even when they're not. Appearances deceive.

But it's not that managers are inherently poor analysts, wishful thinkers, or gullible fools (although some undeniably are). The problem with good ideas is pervasive and systemic. The seductive appeal of destructive ideas is consistent with a wrongheaded global cultural imperative: namely, that ideas and intellect are the dominant source and driving force of wealth creation. Business organizations live in a global innovation economy where this vision has been distorted beyond reason into delusion. The Earth is not flat; the moon isn't made of green cheese; and the future of business innovation doesn't depend on good ideas. It never has.

RAISING KEYNES

Too many businesspeople have been bullied, brainwashed, and highbrow-beaten into believing that economic value creation is rooted in intellectual breakthroughs. They've been persuaded—perhaps even intimidated—by the eloquent specter of John Maynard Keynes's famous cliché: "The ideas of economists and political philosophers, both when they are right and when they are wrong, are more powerful than is commonly understood. *Indeed the world is ruled by little else* [emphasis added]. Practical men, who believe themselves to be quite exempt from any intellectual influence, are usually the slaves of some defunct economist."

Bold words, indeed—but, with respect, what does one expect an elitist Cambridge don with a well-deserved reputation for intellectual snobbery to say? That Great Britain's Industrial Revolution required the collaborative energies of rigorous tinkerers like James Watt and entrepreneurial opportunists like his partner, Matthew Boulton? That the engineering prowess of Isambard Brunel, Richard Trevithick, and Charles Parsons transformed the world's railways, sea transport, and energy production? That John D. Rockefeller's ruthless business practices at Standard Oil altered how ambitious innovators perceived economies of scale and scope? That Henry Ford's relentless focus on design simplicity and production efficiency redefined what manufacturing meant? That college dropouts like Bill Gates, Steve Jobs, Michael Dell, and Mark Zuckerberg might one day lead multibillion-dollar ventures that turned personal computing into mass media worldwide?

Of course not. Keynes and his intellectualizing apostles are above the practicality of all that. Yes, those innovators may have pioneered new industries, overturned establishments, and improved the quality of life for millions. But, in Keynes's reality, they're entrepreneurial 'meat puppets' in thrall to the transcendent influence of defunct economists and dead philosophers. For Keynes, ideas from economics and economists—not tools or technologies from scientists, engineers, and entrepreneurs—are what really rule the world.

Who are we kidding? The arrogance displayed here is exceeded only by its historical inaccuracy. Keynes offers little but the propaganda of his professional genius to make his point. The sweeping pomposity of his assertion recalls George Orwell's tart observation that "some ideas are so absurd that only an intellectual could believe them." Indeed. This particular absurdity thrills intellectuals for an excellent reason: it makes them the heroes and aristocracy of innovation.

That's why understanding this arrogance is important. This intellectual superiority justifies the condescension so many idealists and ideaholics bring to postindustrial innovation. Contemporary Keynesians aren't Harvard-trained economists debating how much money governments should print to stimulate demand. Rather, they're intellectual capital–obsessed ideologues who evangelize that good ideas are more valuable than strong currencies.

Ideas sit at the white-hot center of innovation and value creation in this economic universe. They channel their Lord's fundamentalist ideal, both in their descriptions of how they think the world works and their prescriptions for how they think the world should work. Ideas *über alles.*

This describes the power struggle confronting innovators everywhere. Academics, intellectuals, and the peddlers of good ideas have declared ideas and intellect the central pillars of innovation. People who don't appreciate the power and potential of good ideas are either idiots or apostates. They're doomed. Pity them or damn them, but please get them out of the way.

Coming up with better ideas faster is the last best chance for success in a postindustrial information age filled with knowledge workers competing in a global digitized economy. Got that? Ideas are the precision-guided ammunition, weaponry, and matériel for triumph on postindustrial business battlegrounds. Any dearth of ideas—any dip in intellectual firepower—reveals weakness and invites defeat.

Again, not true. That is warmed-over Keynes digitally enhanced for the Internet age. Organizations and their leaders have overdosed on the idea of ideas. Rethinking the false truisms of good ideas is imperative for smart investment, smarter innovation, and sustainable economic growth.

Businesses wanting to innovate quickly, creatively, and cost-effectively must have the courage to get beyond ideas. Success in global markets requires that entrepreneurs and executives free themselves from this value-destroying dogma. Ideas are the enemy, not because they are intrinsically bad or foolish but because—like glittering objects in front of tiny children—they distract innovators from the challenges that matter most.

THE WEIGHT LOSS FALLACY

You wouldn't measure the brightness of a lightbulb by its smell. You wouldn't assess the acoustics of a concert hall with an electron microscope. You couldn't gauge the resolution of a digital photograph by weighing it. Yet clever executives honestly believe that firms can evaluate the merit of an innovation investment effectively by analyzing its underlying good idea. The cleverest managers are often the most confident in their abilities to judge the value potential of ideas.

The truth, however, is that they can't. Ideas are unhelpful and unhealthy units of analysis for assessing the value of proposed innovations. Most managerial debates around good ideas are as futile as arguing the color of an ounce or asking, "When does music taste good?" This disconnect is fundamental.

Flawed metrics undermine useful meaning. Cost-effective innovation requires measures that put good ideas in their place. These metrics must make it easier for organizations to see how value gets created. An organizational focus on identifying good ideas or better ideas cripples chances for success.

Take weight loss. Many people are significantly and unhealthily overweight. Obesity has become a global public health issue. Multibillion-dollar industries have sprung up to address this challenge. Entrepreneurs from medicine and nutrition successfully compete to procure venture funding for their "solutions."

Fortunately, there's a simple yet powerful idea that medical experts agree can lead to dramatic weight loss for the vast majority of overweight people: "Eat less and exercise more."

"Eat less and exercise more" is the epitome of a good idea. It's cheap, simple, relatively safe, and undeniably effective. A large body of empirical research supports its benefits. The idea is easily understood by virtually everyone who eats and moves. It's remarkably flexible; customization isn't complicated. Different exercises, as well as different foods, may be employed in this approach. Almost every overweight individual who consistently eats less and exercises more can be assured of losing weight. Observable results are achievable within weeks.

Here's the best part: "Eat less and exercise more" is such a good idea that it even scales pretty well. If someone overweight eats *significantly* less and exercises *significantly* more, then they will likely lose *significantly* more weight. As an added bonus, they'll probably look and feel physically healthier, as well. Fantastic.

Like all good ideas, though, "eat less and exercise more" can be harmful if taken to extremes. Eating too little while exercising too much can damage even the healthiest bodies. Success isn't guaranteed, but "eat less and exercise more" enjoys an admirable track record. Few ideas deliver more value for less cost. Yes, side effects like hunger pangs and muscle aches materialize, but they tend to be transitory. The key is that people really are eating measurably less and exercising measurably more.

Of course, this entire example is an exercise in silliness. Everyone—especially the overweight—knows that "eat less and exercise more" is a good idea. But so what? Good ideas have nothing to do with good implementations.

An overwhelming consensus around "eat less and exercise more" exists. But most people will acknowledge that only a small percentage of their overweight colleagues will embrace it seriously. They'd further concede that an even tinier percentage would stick to that good idea longer than three months.

That's the fatal flaw. If everyone knows that the good idea will fail to influence the behavior of the very people who agree they need it most, why is it a good idea? The better argument is that eating less and exercising more may look, feel, and sound like a good idea, but it isn't. It is a truism masquerading as a good idea. For whatever reason, "eat less and exercise more" isn't doable for the vast majority of the world's overweight denizens. It's not a good idea; it's just words.

WHERE THERE'S A WILL, THERE'S AN INNOVATION

This unhappy truth invites the insight of economist Joseph Schumpeter. A brilliant rival to Keynes, he disagreed about the importance of ideas in making innovation happen. "Successful innovation ...," he wrote in 1928, "is a feat not of intellect but of will. Its difficulty consists in the resistance and uncertainties incident to doing what has not been done before. ..."

For Schumpeter, commitment to overcoming resistance and managing uncertainty—not cultivation of intellectual prowess—determines innovation outcomes. Ideas aren't irrelevant to Schumpeter, who was arguably as much an intellectual elitist as Keynes. But he recognized their inherent inferiority to action when it comes to the marketplace. Aspiring entrepreneurs shouldn't invest too much in inferior goods. Real innovators—real leaders—know that actions speak louder than words and behave accordingly.

That's why the idea that good ideas are central to the weight loss challenge is a fallacy. The idea is arguably the least important link in the weight reduction value chain. In theory, any one of dozens of good ideas could lead to the desired outcomes. But in reality, "eat less and exercise more"—or, for that matter, "take diet pills" or "get liposuction" or "have gastric bypass surgery"—matters far less than overcoming individual resistance and managing the uncertainties.

Measuring the actual value of weight loss innovations has less to do with good ideas than how well their users enact them in the real world. Results matter. Innovation is value obtained from actual use, not the promise of potential.

The economic value of innovation is determined more by the quality of implementation than the quality of ideas. In the words of British military historian Michael Howard, "while strategy proposes, it is logistics that ultimately disposes." Decoupling strategic brainstorming of good ideas from the logistics of implementation isn't thinking outside the box; it's self-delusion. Designing the world's most innovative diet and exercise program burns no one's calories merely by dint of its existence.

In other words, if you can't do it, it's not a good idea. If you can't do it well, it's not a good idea. If you can't afford to do it, it's not a good idea. If you can't afford to do it well, it's not a good idea. If you refuse to do it, it's not a good idea. It may be a good idea for someone else; but it is not a good idea for you.

Implementation—*not* the idea—is the superior unit of analysis for assessing value creation. How organizations *enact* ideas—not the ideas themselves—is the soul and substance of innovation. More often than not, implementation ends up redefining both the boundaries and the essence of the original idea. This is reflected in weight loss reality. Seriously, in the harsh light of reality, "eating less and exercising more" has a bigger impact on people's lifestyle than on their weight.

The innovator's struggle to overcome resistance and manage uncertainty almost always reveals deeper insights into the idea. The act of implementation is an act of discovery. Entrepreneurial chefs and software developers depend on this everyday truth. Tasting is different than cooking; executing is different than coding.

The potential of good ideas—however they are defined—overwhelmingly depends on their implementability. Almost anything innovators do that improves implementability almost always improves their capability to wring value from ideas. Quick tastes—quick tests—can make huge differences.

ELEVATION TO PRACTICE

The most obnoxious phrase that lawyers ever inflicted upon innovators is the legal description of invention as *reduction to practice*. The concept that ideas are "reduced to practice" mirrors and mimics Keynesian arrogance. By far the better and more accurate phrase would be *elevation to practice*. Ideas are never reduced by making them actionable and practical; they're elevated and enhanced. Just ask a customer if you don't believe this.

But those words have turned the value upside down. Vocabularies around innovation and invention have been so distorted by neo-Keynesian academics and lawyers that abstract ideas are treated as more valuable than tangible implementations. This may be true for universities and legislatures, but it's demonstrably false for Internets and Walmarts.

SAM WALTON'S MISDIRECTION

Even the world's most entrepreneurial innovators have succumbed to the "good idea" illusion. "I probably have traveled and walked into more variety stores than anybody in America," noted Walmart founder Sam Walton in his 1992 autobiography. "I am just trying to get ideas, any kind of ideas that will help our company. Most of us don't invent ideas. We take the best ideas from someone else."

This isn't false modesty. It's outright misrepresentation. Walton didn't take the best ideas for his company. He took what he saw with his own eyes actually worked. Walton's genius wasn't recognizing his competition's best ideas—it was developing networks of big-box stores that emulated and replicated the best practices that he observed. Walmart's ability to imitate and implement swiftly is what made Sam Walton's variety store surveillance so valuable. The ideas were secondary.

Testing this assertion is straightforward: How valuable would those best ideas have been if Walmart couldn't translate them operationally into "everyday low prices"? Sam Walton's shoppers aren't purchasing ideas—they're reaping the benefits of their successful in-store execution.

Let's go deeper: Much of Walmart's high performance capability comes from superior management of its computing and logistics resources. As Walmart grew into the world's largest retailer, Walton's company also became one of the world's most innovative implementers of information and process technologies.

Two former Walmart chief information officers (CIOs)—each from a different era of the company's success—separately emphasized that thinking big was never the issue. Walmart's people had almost as many innovative ideas as its superstores had products. The managerial challenge had nothing to do with either the quality or quantity of ideas. Faster, better, and cheaper testing and implementation—that is, everyday, low-priced testing and process implementation—was the dominant managerial concern.

Walmart's biggest partners—from Procter & Gamble (P&G) to Colgate-Palmolive to Unilever—also generated excellent ideas, recalled one former CIO. But the nagging concern remained: Could those partners implement systems and processes for Walmart that consistently and reliably delivered? "We're always looking for better ways to do things," she observed, "but finding a better way to do something is not the same as being able to do it every day of the year."

Walmart didn't just want partners to present clever proposals. The Bentonville behemoth insisted that new capabilities be developed. Creating new operational capabilities made implementing the next round of ideas cheaper and easier, thus creating a virtuous cycle for growth.

To their credit, much of Walmart's leadership understood that effective implementation meant going beyond detailed planning, good communications, and comprehensive training. Before Walmart's technologists would write software to make a business process more efficient, the company actually had its programmers perform the jobs they sought to automate.

What was the breakthrough behavior? Walmart baked implementability into its internal innovation process.

As former Walmart CIO and Microsoft COO Kevin Turner observed, insisting developers actually perform the tasks they coded for dramatically improved software quality and usability. They were pushed to address internal resistance and uncertainties that went well beyond the technical challenges the business process innovators anticipated.

"We found requiring our programmers to do those jobs gave us a very good return," said Turner. Software built on real-world behaviors rather than reported requirements reduced implementation costs and risks.

Its founder was infatuated with ideas, but Walmart's innovation culture emphasized the virtue and value of implementation. Walmart's technical and organizational investments in implementability made good ideas economically actionable.

The firm's ongoing enterprise success results from rigorous attention to ROI, which in this case means "return on implementation," not just "return on investment." Everyday implementations—not the best ideas—are what make Walmart Walmart.

FROM VERBAL TO VERB

Walmart's observed experience is consistent with my own. I've never visited a firm that—with a little poking, prodding, and play—couldn't come up with more good ideas than it could handle.

By contrast, I've rarely been to organizations where implementing good ideas is simple and straightforward. Implementation frictions typically burn out and blister the inspiration of ideas. Confronting, seducing, and overcoming the coefficients of implementation friction pose the real innovator's dilemma.

The dysfunctional disconnect between ideas and their implementation afflicts innovation initiatives worldwide. Many firms have run enterprise campaigns that solicit good ideas and suggestions for innovation and improvement. They're pushing for greater openness and participation. They want empowered employees. They see management thrilled by the breadth and variety of ideas.

Rubbish. The disappointments and frustrations that stem from this tactic deeply wound. Management ends up unhappy with the scattershot quality of the suggestions. Employees grow even more cynical because, six months later, barely a handful of their best ideas are being tested seriously (if at all). So call in the external consultants to make up for those internal inefficiencies. (I should know; I'm one of them!) It's unfair, unjust, and unproductive.

This pattern is pathological. Management isn't looking to test or implement suggestions immediately; it wants to study and analyze them first. Then it wants to review the analysis before proceeding.

Analysis is culturally cherished over action. There's always—*always*—good reason to spend a little more time exploring the ideas. This is particularly true for ideas that aren't designed to be tested quickly, cheaply, or easily. When the analysis concludes, it gets analyzed, too. Call it "quality assurance."

This betrays the bias. Look at most corporate initiatives eliciting ideas, improvements, and suggestions. They're rarely structured in forms that make taking the next step clear or straightforward. Initiatives intended to energize creativity and innovation end up having the opposite effect: they remind everyone why it's so difficult to try anything new.

WHAT SHOULD WE DO?

When aspiring innovators with a good idea say, "We should do X," the most important word in that sentence isn't the *X,* but the *do.* What does it mean, in practical terms, to "do" X? What explicit actions to people take? What specific choices must they make? These aren't tactical questions; they're how X will come to life.

When a firm's innovation leadership declares, "We need to explore Y," the most important word is *explore.* When discussing innovation's next steps, verbs always matter more than objects. The verbs determine how—or if—that friction is overcome.

Successful innovators invest in actions that constructively bridge the gaps between conceptual ideas and practical implementations. Like Walmart's leadership, they're constantly looking for ways to translate good ideas quickly and cheaply into forms and formats where they can be tested, likewise quickly and cheaply. Identifying those actions—selecting those forms—is how ideas are elevated to practice.

When James Watt came up with the idea that a separate condenser might revolutionize the efficiency of the Newcomen steam engine, his immediate actions proved more revealing than his insight. The scientific instrument maker quickly built models of his proposed machine—a brass surgical syringe serving as a piston pump for his rapid prototype. His idea meant little until Watt gave it tangible form. He didn't play with the idea of a separate condenser; he experimented with actual models to test and refine his practical hypotheses about power.

Similarly, Apple cofounder Steve Jobs had strongly held ideas about what personal computing should mean when he first saw the Alto workstation prototypes at Xerox's Palo Alto Research Center in the 1980s. His genius was similar to Sam Walton's. He viscerally grasped that interfaces and hardware/software integration could be produced faster, better, and far cheaper than Xerox ever dreamed. He was right.

With its first- and second-generation Macintosh machines, Apple's superior pop implementations of technological ideas that originated with others redefined the nascent industry's innovation expectations. Apple's superior abilities in prototyping designs and designing prototypes allowed the company to tightly link conceptual breakthroughs with technical implementations.

A PROTOTYPE IS A HYPOTHESIS

Conventional wisdom sees prototypes as models; selective slivers of possible reality. Technology historian Thomas Hughes once described them as slices of congealed culture. All that is true.

But every prototype is also a hypothesis. That is, a *prototype* is a proposition explicitly designed to explain—or make an educated guess about—an observable phenomenon.

The dictionary definition of the word *hypothesis* suggests as much:

1. a proposition, or set of propositions, set forth as an explanation for the occurrence of some specified group of phenomena, either asserted merely as a provisional conjecture to guide investigation (working hypothesis) or accepted as highly probable in the light of established facts.

2. a proposition assumed as a premise in an argument.

3. the antecedent of a conditional proposition.

4. a mere assumption or guess

"A prototype is a hypothesis" means that prototypes are educated guesses about the future—the future of how the prototype might perform, the future of how potential users might react to it, the future of how it might be produced or manufactured, the future of how people might sell or market it, the future of how researchers might explore and test its technical features and functionalities further, the future of how designers might shape or refine its look further, etc. A prototype describes a potential future worth testing.

A prototype's design hypothesis is an assertion of how a design choice creates value. To wit, "What option do you like more?" "How do we design this better?" and the like. The prototype embodies the design hypotheses to be tested.

An iPad prototype was an excruciatingly well-crafted hypothesis about the future of digital media consumption. An Android phone prototype was a superbly educated guess about mobile communications. No doubt, future prototypes of the Toyota Prius will set forth important propositions about the future of four-wheeled carbon-neutral transport.

In other words, prototypes inherently embody the propositions, antecedents, assumptions, and guesses that define a hypothesis. Crafting a prototype—digital, physical, or virtual—means crafting a hypothesis. If there isn't a hypothesis, then it's not a prototype. The more refined the hypothesis, the more refined the prototype; the more speculative the hypothesis, the more speculative the prototype.

But a hypothesis for whom? Does the hypothesis or prototype have the user at the top of the mind? The manufacturing team? The interface designer? The accountants? The salespeople? Or all of them?

Who gets to formulate the hypothesis? Who owns it? How will it be tested? These simple and straightforward questions quickly surface the cultural and organizational tensions confronting all design communities. Are the hypotheses rooted in market, technical, or aesthetic concerns? Are the hypotheses managed as an integrated portfolio of educated guesses? Or is each hypothesis the equivalent of a requirement or spec demanding confirmation and compliance?

But like any physics, chemistry, or biology experiment, the prototype is meant to do more than articulate and test a hypothesis. It is also intended to persuade. A successful prototype—like a successful experimental hypothesis—effectively persuades people of its value and correctness. A prototype that doesn't persuade—like an experiment that doesn't support its hypothesis—is seen as a failure. Successful prototypes win in the marketplace of ideas.

For Amazon, a few dozen lines of code forged the innovation link between actionable idea and real-world testing. The ultimate outcome was a billion-dollar transformation of the Internet giant's business model. Greg Linden was part of a small group of Amazon intrapreneurs (people who act like entrepreneurs within a large company) who believed that the retailer could create an even better shopping experience for customers if its software could make intelligent recommendations on what to buy.

"I hacked up a prototype," said Linden. "On a test site, I modified the Amazon.com shopping cart page to recommend other items you might enjoy adding to your cart ..."

Linden's little experiment quickly scaled into huge success. But his simple hack is what gave the recommendation engine its power and influence, both inside and outside Amazon's virtual walls. The test results—undeniable evidence that people were persuasively nudged to click on the recommendations and add to their shopping carts—were compelling. The prototype enacted the idea. The experiment made that potential real.

What these examples have in common—besides success—is recognition that economic value emerged not from ideas but their interactional expression. Words are not pictures are not prototypes. Initial expressions of ideas powerfully influence their ultimate implementation.

Watt's brass syringe condenser and Linden's recommendations hack are misunderstood as ideas or inspirations. They aren't. Rather, they are expressive experiments. Expressiveness gives the idea form and context; experiments test form and context against an expectation.

For Watt, that expectation—that hypothesis—was a radically new engine for efficient power generation; for Amazon, it was a disruptively innovative recommendation engine for generating customer participation and purchase. Aligning expressive experiments with expectations, or hypotheses, is what successful innovators do. Less successful innovators have less expressive experiments. They're too fond of their ideas.

Look carefully at the history of technology, entrepreneurship, or business innovation. A persistent pattern emerges. Successful innovators talk about ideas, but they invest their time, money, and ingenuity in expressive experimentation

Their competitive success comes from getting more value faster from expressive experimentation. The economics of this type of experimentation play an enormous role in identifying and influencing the economics of implementation.

For organizations looking to innovate, the message is simple and stark: improving ideas means little if the ideas' expressiveness is not improved too. Expressiveness—the forms and flexibilities of expression—is the "secret sauce" to innovation success.

Getting superior returns from expressiveness, however, depends on how well expressive experimentation helps implementability. Quickly creating the portfolios of expressive experiments that offer the greatest innovation insights with the most acceptably manageable risk.

WHAT'S THE BIG EUPHEMISM?

Rhetoric matters—but actions still speak louder than words. The unhealthiest innovation conversations that I hear tend to revolve around the question: "What's the big idea?"

Cultures and leaderships of organizations emphasizing big ideas want to transform their business. The focus of their time and energy is vetting and analyzing these big ideas to see if they are worthy of further study and investment. But, as the Sam Walton vignettes from earlier reveal, this focus is futile. "Ideas" have degenerated into a conveniently lazy verbal shorthand that obscures meaning rather than promotes understanding.

A top executive who challenged people by asking, "What's the big notion?" or "What's the big intuition?" or "What's the big belief?" would likely inspire eye-rolls and snickers. But because ideas are valued as a politically correct enterprise brand, people feel compelled to treat such questions seriously.

They shouldn't. The next chapter explores and defines what makes a good investment in expressive experimentation.

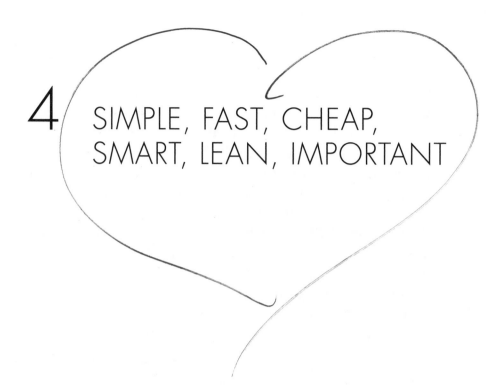

4 / SIMPLE, FAST, CHEAP, SMART, LEAN, IMPORTANT

Successful innovators know that six ingredients are essential to compete profitably in global markets: simple, fast, cheap, smart, lean, important. Google, Facebook, and Amazon understand this. So do McDonald's, Tesco, and Walmart. They know how to mix their ingredients well. That's why they win.

This book explains how successful experimentation drives successful innovation. But successful doesn't mean expensive, complicated, or sophisticated—quite the opposite.

Simple, fast, cheap, smart, lean, and important experiments can supercharge any serious innovation process. This book offers market-tested recipes for whipping up fast, cheap, and simple business experiments that transform value creation. Faster, cheaper, and simpler experiments mean faster, cheaper, and simpler innovation. Experiment better; innovate better; win better. That's the promise.

The methodology is powerful, practical, and proven. It works. Your organization can use it formally as a key innovation process. You can use it to collaborate informally with colleagues to innovate.

It's flexible, scalable, and robust. No revolution(s) required. No special mathematical, scientific, or statistical expertise is necessary. But results require more than just talk.

The best innovators typically run the best experiments. The best experiments are designed to generate the greatest opportunities for insight in the least amount of time. Amazon's transcendentally successful recommendation engines, for example, weren't born of strategic planning. They rapidly evolved from unauthorized experiments that looked at how shoppers might respond to unsolicited suggestions onscreen. Simple technical tests launched a profitable multibillion-dollar business platform.

The best business innovators typically run the best business experiments. Their experiments seek to create new value, both for their customers and themselves. Business experimentation is about identifying economic value; rapid business experimentation is about quickly identifying economic value for cultivation.

This was as true at the dawn of the Industrial Revolution as it was when the automobile and aviation industries first began. People forget—or don't realize—that James Watt's steam engine was a spin-off from a scientific model that he was fixing for Glasgow University; that the Wright brothers pioneered wind tunnel experiments for airplane design; and that Henry Ford's factories were as much research laboratories for process efficiency as mass production facilities for cars. Experimentation literally and figuratively drove innovation.

This value experimentation ethos also holds true for the semiconductor and software revolutions. Iterative experimentation drives the Web 2.0 postindustrial surge of interactive and collaborative innovation.

Rapid experimentation for business innovation has been a cultural core competence of market leaders worldwide. The Googles, Walmarts, Apples, Amazons, Tescos and Toyotas understand this. They live it. They've internalized it. They hire for it. They invest in it. Their best customers expect it. These firms undeniably possess their own innovation cultures and experimental styles. But so what? Important similarities transcend these distinctions. The most important cultural and organizational common cause by far: Actions speak louder than words.

That's the secret to healthy and vibrant innovation cultures. Innovation isn't a managerial seminar or analytic exercise. Innovators act. They do. They test. They respect results. They go beyond throwing a prototype against a wall to see if it sticks.

Successful innovators don't just celebrate the innovations they offer; they internally prize the experiments and experimentation that made innovation possible. They acknowledge the key experiments that commanded attention, respect, and action. Google founders Larry Page and Sergey Brin point to their firm's data-driven experimentation as integral to its innovation prowess.

In that respect, Google's message is indistinguishable from that of Walmart or Toyota or McDonald's. What innovators do matters more than what they say. In other words, they don't "act smart"; they perform smart acts. The distinction is enormous. Acting smart is a performance; smart acts deliver performance. Serious innovators strive to deliver good performance fast. Rapid experimentation invites rapid action.

Experiments are actions that go beyond words. A good business experiment speaks more eloquently than a good business idea. Good experiments are more persuasive, more useful, and more valuable than good ideas. Good experiments manage expectations better than good ideas.

Experienced innovators know this. They understand that good ideas aren't good ideas unless they can be turned into good experiments. Experimental outcomes can't be chatted, charmed, or fast-talked into existence. You have to do it. Experimentation demands action. Playing with ideas becomes indistinguishable from performing quick experiments. "Leading by experiment" becomes a way to lead by example.

Why do so many organizations have so much trouble embracing faster, simpler, and cheaper experimentation? The answer is as provocative as a good hypothesis. Experimentation is undervalued

WE ARE POOR AT ASSESSING THE VALUE OF IDEAS

Features are built because teams believe that they are useful, yet in many domains, most ideas fail to improve key metrics. Only one-third of the ideas tested at Microsoft improved the metrics that they were designed to improve. Success is even harder to find in well-optimized domains like Bing. Jim Manzi, head of Applied Predictive Technologies (APT), wrote that at Google, only "about 10 percent of these [controlled experiments, were] leading to business changes." Former Google executive Avinash Kaushik wrote in his Experimentation and Testing primer that "80 percent of the time you/we are wrong about what a customer wants." Mike Moran wrote that Netflix considers 90 percent of what they try to be wrong. Regis Hadiaris from Quicken Loans wrote that "in the five years I've been running tests, I'm only about as correct in guessing the results as a major league baseball player is in hitting the ball. That's right—I've been doing this for five years, and I can only "guess" the outcome of a test about 33 percent of the time!" Etsy's Dan McKinley observed that "nearly everything fails" and "it's been humbling to realize how rare it is for them [features] to succeed on the first attempt. I strongly suspect that this experience is universal, but it is not universally recognized or acknowledged." Not every domain has such poor statistics, but most people who have run controlled experiments with customer-facing websites and applications have experienced this humbling reality: we are poor at assessing the value of ideas.

Ron Kohavi, Alex Deng, Brian Frasca, Toby Walker, Ya Xu, and Nils Pohlmann, "Online Controlled Experiments at Large Scale" (http://www.exp-platform.com/Documents/2013%20controlled ExperimentsAtScale.pdf).

and unappreciated wherever words speak louder than actions. Period. The proliferation of Microsoft PowerPoint presentations is inversely proportional to a propensity to experiment.

If your organization consistently prefers sophisticated analysis to simple experimentation, you're innovation-disadvantaged. If your organization would rather run twenty comprehensive experiments next month than two simple experiments tomorrow, you're destined for innovation dysfunction. If your firm believes that its most gifted strategic thinkers will generate its most provocative experiments, you're competitively crippled in the innovation race. These pathologies pervert productive innovation performance.

Stop. Reassess. The economics of experimentation are your friend. Marginal investments in simple and quick business experiments can reap wildly disproportionate returns. No revolution required. This needn't be difficult, time-consuming or expensive. It can—and should—be cheap and easy.

But a willingness to ask simple questions and the discipline to take tiny but quick steps to answer them is essential. Even one-shot commitments to simple questions and quick steps can get you more than halfway to high-impact innovation. Breakthrough experiments—the ones that make eyes widen and jaws drop—materialize when colleagues and customers collaborate to ask those questions and take those steps.

Ride the rapid experimentation learning curve. You'll be energized and exhilarated. The more you experiment, the easier it becomes. The easier it becomes, the more you experiment. Virtuous cycles are wonderful. If rapid experimentation doesn't become easier the more you and your colleagues do it, then you're not doing it right. When you're doing it right, the results are remarkable. You can't help but do well.

Are you willing to win big by spending small? Are you prepared to leverage large insights from small experiments? Are you comfortable taking simple, quick and tiny steps to "winnovate"?

Do you believe that actions speak louder than words? Does your organization agree? Attitude is as essential as aptitude. If you prefer complexity, giant leaps, and huge budgets, this methodology isn't for you. Earning great gain without great pain is possible. Breakthrough innovation doesn't have to hurt. Be open. You don't have to change your mind—but you do need to tweak your mindset.

Consider this high-impact example. A modestly successful business simulation company relied heavily on its website for its marketing. Although the site made sampling the simulations simple and friendly, sales proved disappointingly low; and this surprised management. Customers rated the simulations highly; prospects who spent time on the site playing with the product tended to buy. Both the sales and qualified lead numbers should have been stronger. A costly site redesign failed to improve the outcomes.

A simple experiment identified a serious problem: Visitors who played the on-site simulations had every intention of purchasing. But growing numbers of them simply abandoned the site rather than register their email address before sampling. The company's marketers had two good reasons for wanting those addresses: data gathering and deterring unqualified prospects. Any serious prospect willing to register was welcome; junk samplers who just wanted to play the simulations were not.

The impeccably credentialed marketing team debated whether focus groups or web surveys would best diagnose their site's underperformance. But this dithering proved unbearable for the impatient company cofounder. He and the firm's webmaster picked a third option.

They designed a quick experiment. Its hypothesis? Marketing's lust for immediate site registration scared off desirable prospects. The experimental design? Half of all visitors would be asked randomly to give their email address immediately; the rest would be prompted to share their address only if they sampled a simulation for longer than two minutes.

This experiment, which took less than two days to launch, produced statistically stunning results. Within a week, the data revealed that over half the hundreds of visitors asked to register instantly abandoned the site. By contrast, almost all samplers who played for more than two minutes provided their emails when asked—and roughly 40 percent of them were precisely the kind of qualified prospects that the marketers most desired.

User actions also speak louder than words. The experiment revealed that pushy requests alienated perhaps a third of the firm's most desirable prospects. Those findings transformed the site's sales positioning and marketing purpose. Listening to users frequently reveals less than innovatively observing them.

"That experiment directly led to more prospects, more revenue, and more profit faster and cheaper than anything we had ever done," said the company's cofounder. "It was low-hanging fruit."

Low-hanging fruit tastes delicious. The experiment's impact went well beyond productivity and profit. The firm's innovation culture shifted. Simple, cheap experiments became part of the managerial process. If a quick experiment would test a key business assumption or enhance an MBA-driven analysis usefully, then just do it. "Why experiment?" became "Why *not* experiment?" That inversion lies at the heart of cultural transformation.

Simplicity, speed, and smarts don't need new technologies or expensive instrumentation to challenge assumptions and transform established fields. Shockingly simple and cheap experiments can—and do—win Nobel prizes. Having the humility to insist that big thinking must act small is key. Huge ideas don't inherently require huge experiments; on the contrary, tiny ones can pack an enormous punch.

Pioneering research by psychologists Daniel Kahneman and Amos Tversky in the 1970s challenged fundamental economic beliefs that people were rational decision makers. Their study argued that most people couldn't assess decision information well. In reality, most people are poor intuitive statisticians who consistently miscalculate the odds and misunderstand probabilities. Even experts—the top people in their field—consistently make simple but significant mistakes.

Kahneman and Tversky had observed profound decision-making pathologies in fields ranging from the medical to the military. But empirical observation wasn't enough for them. They wanted persuasive experimental proof. So they designed simple pencil-and-paper questions for mathematical psychologists—experts in both psychology and statistics—to answer at a conference.

The results—which have been replicated across cultures and disciplines—couldn't have been better if they had been made up. Disturbingly large proportions of professors and PhDs gave exactly the wrong answers. Experts succumb to the same cognitive biases and pathologies as ordinary people do. Human minds—even the best-educated ones—are riddled with vulnerabilities and flaws that reliably undermine rational calculation.

Behavioral economist Dan Ariely describes this dysfunctional dynamic in his best-selling book *Predictably Irrational: The Hidden Forces That Shape Our Decisions*. Rational behavior is trickier and more elusive than most classical economists once thought. Behavioral economics insights have simultaneously undermined and redefined large chunks of economic theory.

These simple experiments led directly to Kahneman receiving the 2002 Nobel Prize in Economics (Tversky had died in 1996; the prize is not awarded posthumously). Behavioral economics, the discipline that they effectively launched, has profoundly influenced not only public policy research, but the epistemic importance of experiment in economics.

The field's global reach is impressive. Smart investors and savvy financial executives increasingly look to behavioral finance insights to inform their own capital allocations. Behavioral economists, like Ariely and the University of Chicago's Richard Thaler, continue to build on Kahneman and Tversky's design tradition of fast and cheap experimentation.

Could the original experiments have been more extensive and expensive? Of course. Experiments can always be more complicated and cost more. But much of the duo's Nobel prize–winning genius was demonstrating conclusively that experimental cost and complexity were unnecessary.

Simpler, cheaper, and faster experimentation invited—indeed, inspired—imitation, emulation, and replication. The discount economies of behavioral economics experimentation accelerated its growth and influence. Year in and year out, elegantly simple experiments yielded disproportionately useful results. You don't need multibillion-dollar Large Hadron Colliders to effect scientific revolutions. Tiny can be terrific.

Quickly leveraging simple experimental design also can help manage innovation complexity. In the late 1970s, General Mills, the Fortune 500 food company, created a new snack food—Fruit Roll-Ups—with an unusual texture and form factor. Successful production required novel machinery and new processes.

Unfortunately, the production line couldn't deliver. Quality and quantity were both inadequate. Spot fixes failed. Tweaking the machinery only made things worse. The snack's creators couldn't figure out what was going wrong. The equipment manufacturers proved unhelpful. Expensive engineering experts parachuted in from headquarters were stumped. The machinery, process, and the snack food itself were too novel for conventional process improvement analysis and repair.

Some of the powers-that-be thought the solution was to shut down the line—at the potential cost of millions of dollars. The situation was so desperate that management gave up on a quick fix. The food giant tasked an in-house statistician to analyze the failed repair efforts in hopes of finding the fatal flaw. But the statistician did no such thing, because he knew that he brought no meaningful production expertise to the data that he would see.

Instead, he flew to the plant. He gathered the chemists, production engineers, and line operators into a room. With flip charts at the ready, he asked painfully simple questions. How should the machines work? How are process failures monitored? What failures surprised you? What data would help make possible better decisions about reconfiguring a particular machine or production process? Do people agree on data interpretation?

Not even two days later, this intense facilitated discussion yielded a rough consensus. The agreement didn't identify what was going wrong; rather, it centered on the variables most likely responsible for ruining the production runs. Guided by their statistical interlocutor, the team quickly devised a set of rapid experiments to test their hypotheses. A bit of math calculated the fastest sequence for running the experiments. Running them took two more days.

The result? These runs revealed hidden and unanticipated process interdependencies. The evidence was unambiguous. From that moment on, fixing the problems proved straightforward. Tinkering with individual machines or isolated subprocesses would have been futile.

So what happened? The line was running at capacity within a week. But—almost as important—the entire production crew now had visibility into the entire Fruit Roll-Ups production process. The experiments had gone beyond identifying and repairing a broken process. They dramatically informed and improved the on-site technical capabilities and competence of the people involved. Rapidly designing and implementing experiments made people smarter—and making them smarter faster was a big win.

Yes, the team succeeded where the experts had failed. But understanding the true nature of experiment-enabled success is essential.

The central business insight is that the improvised factory team did not 'solve the problem.' The team succeeded by designing experiments that made solving the problem possible. Experiments were the essential means to the productive end. The solution couldn't have materialized quickly or cheaply without the experiments. Experiments didn't solve the problem, but proved the fastest, cheapest, and simplest way to determine how to solve the problem.

This was the big win. The industrial statistician didn't succeed by imposing his quantitative expertise upon a novel production process that he didn't understand. He succeeded by tapping the collective knowledge of the group to facilitate effective experimental design. This distinction isn't subtle; it's enormous. By focusing on fundamentals first, fast, and foremost, participatory success became possible.

The conclusion appears almost counterintuitive: Don't invest in business experiments to solve problems or create opportunities. Invest in experiments to quickly, cheaply, and easily gain insight into solving problems and exploiting opportunities. The best experiments purchase actionable insights at fantastic discounts, and with remarkable speed. They produce bargain breakthroughs. They deliver great value for money—if they are allowed to do so.

These simple stories have similar themes. Perhaps the most important is that each of these experiments addressed a fundamental question that everyone agreed mattered. No ambiguity, no confusion, no doubt. Actionable insights into these fundamental questions were crucial.

Just as important, these fundamental questions could be framed simply and clearly. Everyone understood them. Everyone understood the reasons for the experiments. Everyone understood the experimental outcomes. Simple, fundamental questions lead to simple, fundamental experiments that suggested compelling and actionable next steps.

Cost was the next central theme. None of these experiments proved expensive to design, implement, or assess. On the contrary, they were cheap. They were cheap on an absolute basis and—significantly—they were cheap relative to the analytical or advisory alternatives.

More precisely, the organization saw the costs of the experiments as too tiny to matter. The cost-benefit ratio was so overwhelmingly favorable that doing the experiment made more sense than not doing it. When money matters less, people and potential matter more.

The value of speed and immediacy can't be overstated. These experiments were designed with urgency in mind. Speed and quickness make simplicity and cost even more attractive. (Simple, cheap, and slow is rarely a winning combination in competitive enterprise.) Faster is better. Speed accelerates value creation and intensifies value appreciation. And speed means that more can get done in less time.

Google offers a brilliant example. When I ask people what kind of company Google is, the overwhelming majority responds: It is a search company. I tell them they're wrong and ask a simple question to prove it:

If Google could improve its search quality by an order of magnitude—10 times by whatever definition of better you value most—but you'd have to wait one hundred seconds for the results, do you think Google would become a more valuable company or less? Without hesitation, the overwhelming majority of respondents agree: "Less ... much less."

The point is that Google isn't a search company—it's an instant search company. Google's perceived and actual value is directly related to its search speed. (Indeed, Google implicitly brands itself this way by posting onscreen the time that it takes to perform a search.) Similarly, McDonald's isn't just a food company—it's a fast-food company. Customers grow impatient if the lines are too long or their orders don't come fast enough. Speed of service is intrinsically part of their core value proposition.

"Speed of experiment" is an intrinsic ingredient of this methodology, which ideally would make every organization a Google or a McDonald's of business experimenting. The methodology reflects British radar pioneer Sir Robert Watson-Watt's "Law of the Third Best." Watson-Watt, who led Great Britain's development of radar systems in the 1930s in anticipation of World War II, famously argued, "Give them the third best to go on with. The second best comes too late; the best never comes."

Watson-Watt's emphasis on faster being better than good enough made him an experimentalist and innovator par excellence. He was the point man for British defense electronics. Don't overlook or underplay the brilliant bargain that Watson-Watt's law strikes between simple experimentation and grand ambition.

Google has extraordinarily grand ambition and vision. So do Walmart and McDonald's. Kahneman and Tversky deliberately sought to topple the foundations of an established academic field. The new product ambition at General Mills may not have been grand, but its innovation situation was demonstrably in desperate straits.

Great vision and grand ambition did not lead to ever greater and grander experimentation. On the contrary, great vision and grand ambition guided rapid-fire investment in faster, simpler, and cheaper experiments. This may seem paradoxical, but the more compelling the organizational vision, the faster and easier it becomes to devise simple and fundamental experiments. That's a central 5×5 insight and organizing principle.

The dangerous and destructive enterprise myth is that experiments need to be as grandiose and comprehensive as the company's vision. Nothing could be further from the truth. The best way to enrich, enhance, and advance the boldest ambitions is through the simplest, fastest, cheapest, and most fundamental experiments. This book explains how.

5 THE FUNDAMENTAL VALUE INNOVATION: STEALING FROM WARREN BUFFETT

My best professional perk as a young *Washington Post* reporter was the opportunity to chat with Warren Buffett. The Sage of Omaha—already celebrated as history's savviest stock picker—owned a big chunk of the Washington Post Co. and served on its board. His great wealth aside, Buffett is likeable, quotable, and funny. He is the rich uncle that you wish you had.

We talked a lot about his deals. I didn't know then that those conversations would be less important as quests for quotes than as tutorials about how the world's best investor analyzed value. Buffett's real-world choices sharply contradicted cherished academic truisms about efficient markets and innovation. His investment philosophy differed markedly from financial axioms then taught at Harvard, Stanford, and MIT. His bottom-line results, of course, were even more impressive than his rhetoric.

In terms of insight and influence, Buffett has proved to be a multibillionaire Peter Drucker. His business advice is as pithy and profound; his strategic sensibilities are as sharp. They endure. The

simplest proof is that Buffett and Drucker can be reread profitably decades later. No serious investor or senior executive reads
Buffett's Berkshire Hathaway "shareholder letters" without feeling
pricks of self-consciousness about his or her own investment and
leadership styles:

- Price is what you pay. Value is what you get.

- … we try to stick to businesses we believe we understand … If a
business is complex or subject to constant change, we're not smart
enough to predict future cash flows. Incidentally, that shortcoming
doesn't bother us. What counts for most people in investing is not
how much they know, but rather how realistically they define what
they don't know. An investor needs to do very few things right as
long as he or she avoids big mistakes.

- Growth benefits investors only when the business in point can
invest at incremental returns that are enticing—in other words, only
when each dollar used to finance the growth creates over a dollar of
long-term market value.

- … I would rather be certain of a good result than hopeful of a
great one.

- My most surprising discovery: the overwhelming importance
in business of an unseen force that we might call "the institutional
imperative." … For example: (1) As if governed by Newton's First
Law of Motion, an institution will resist any change in its current
direction; (2) Just as work expands to fill available time, corporate
projects or acquisitions will materialize to soak up available funds;
(3) Any business craving of the leader, however, foolish, will quickly
be supported by detailed rate-of-return and strategic studies prepared
by his troops; and (4) The behavior of peer companies, whether they
are expanding, acquiring, setting executive compensation, or whatever, will be mindlessly imitated.

- You don't have to be an expert on every company, or even many.
You only have to be able to evaluate companies within your circle of

competence. The size of that circle is not very important; knowing its boundaries, however, is vital.

- In the final chapter of *The Intelligent Investor*, Ben Graham says: "Confronted with a challenge to distill the secret of sound investment into three words, we venture the motto, 'Margin of Safety.' [More than four decades] after reading that, I still think those are the right three words."

The discipline, rigor, and wit that Buffett brought to vivisecting value transformed my innovation assumptions. Sophisticated technical analyses were overrated. Perhaps the best innovation insights come from asking simpler, more Buffett-like questions. Buffett constantly revisited his financial fundamentals; maybe innovation fundamentals require regular revisitation as well. Perhaps the devil isn't in complex details—maybe he lurks within misunderstandings and mischaracterizations of fundamental value.

Buffett's investment rhetoric seemed ideal for reevaluating strategic innovation. Practically every aphorism, epigram, analysis, anecdote, and insight that he assigned to value investing is reenergized when defining innovation. What are circles of innovation competence? What margins of safety serve innovation investors best? What institutional imperatives distort innovation investment? What gaps persist between the price paid for innovation and the value obtained?

If Warren Buffett invested in innovation as an asset the same way that he invests in corporations as assets, then WWWBD— namely, "What Would Warren Buffett Do?" The business challenge is bringing a fundamental value sensibility to innovation investment.

The most important epiphany emerged from a "Buffettism" that he employs in private conversation and public talks. He likes pithy and accessible one-liners to explain himself, and he found one for his investment philosophy: "The simplest definition of value investing is that it's about how do you buy a dollar for fifty cents."

Wow. How do you buy a dollar for fifty cents? What a great question. Who doesn't want to buy a dollar for fifty cents? Why shouldn't savvy investors look for fifty-cent dollars? The concept is instantly appealing. Everyone—from the C-suite to middle management to the production line to the customer contact center—gets it.

"Good artists borrow, great artists steal." In that spirit, I shamelessly ripped off Buffett's value-investing heuristic and adapted it for my own innovation research. Traditional financial investors emphasize value investing; business innovators should explore value innovating. How do you buy a dollar's worth of innovation for fifty cents? That's how fundamental value innovators should assess their innovation investment opportunities.

What does a dollar's worth of innovation mean to individuals and institutions? What tools, techniques, and technologies can help entrepreneurs and executives buy a dollar's worth of innovation for fifty cents, a quarter ... a dime ... or maybe even less? Being disruptive or more innovative isn't enough. A sustainable business investment case is needed. If Buffett is to be believed, the best answers to those questions identify the best innovation investments.

Anyone should be able to buy a dollar's worth of innovation for a dollar. But that's not an investment; it's an exchange. Purchasing a dollar's worth of innovation for 85 or 90 cents is better. But that cheats Buffett's and Graham's vital "margin of safety" concept. Fundamental value investors avoid overpaying for stocks; fundamental value innovators shouldn't overpay for innovation.

But that's where investor psychology distorts fundamental value perceptions. Many organizations believe that innovation is so important that they're willing to pay a premium to get it. Innovation is their public path to growth and profit. But when the final accounting is done and the returns tallied up, they frequently find that they've invested two, three, or even four dollars for a dollar's worth of innovation. In this scenario, innovation is less a source of economic value creation than an enterprise loss leader. Their financial return on innovation (ROI) is less than zero.

This pathology recalls a cruel business joke that pops up in Silicon Valley, venture capital, and Harvard MBA circles:

Q: *What do you call an investment that takes two dollars to produce a dollar's worth of value?*

A: *Strategic.*

That punch line provokes snorts of cynical laughter from consultants and their now (slightly) less well-heeled clients. Their ex-clients find the joke less amusing. (Buffett, on the other hand, enjoyed it.)

The punch line's point is that strategic innovation shouldn't be a loss leader. Strategic innovation should create economic wealth, not destroy it. As Buffett observed, "Price is what you pay; value is what you get." Value innovation is about getting value comfortably above and beyond the price paid. Innovation that undermines enterprise value creation is a failure.

Steve Jobs was a strategic value innovator. The numbers—not just his company's products—say so. Apple Computer spent far less on research and development (R&D) than its competitors during its recent decade of rapid growth. Despite the competition investing more R&D dollars on both an absolute and a percent of sales basis, Apple enjoyed significantly greater innovation returns. Did Apple have its innovation failures? Of course. But none were spectacularly expensive. Steve Jobs's company carved out remarkable margins of safety in its innovation investment.

How was that possible? Focus. Apple focused on its interface design and form factors, as well as facilitating elegantly simple and simply elegant user experiences. Lower prices matter less than higher perceived value. The firm both understood and coolly expanded its innovation circles of competence.

Technical excellence did not require financial largesse. Both technologically and financially, Apple's innovation investment outperformed rivals Nokia, Microsoft, Palm, Research in Motion (the maker of BlackBerry devices) and other peer competitors.

Other innovation leaders like Tesco and Amazon display similar investment behaviors. Paying a premium for innovation cuts against their cultural grain. Their investment philosophy sees value for money—not disruption or breakthroughs or "blue ocean"—to set innovation expectations.

In reality, their innovations do disrupt and break through. Value innovating doesn't preclude paradigm shifts. For example, Tesco's loyalty programs and Amazon's recommendation engines—like Apple's interfaces and apps—fundamentally reshaped the competitive landscapes of their industries. But, first and foremost, they delivered measurable value-to-cost/value for money to key customers. Boldness appeared everywhere but in the budget.

"I think frugality drives innovation …," observed Amazon founder and chairman Jeff Bezos in an interview. "When we were [first] trying to acquire customers, we didn't have money to spend on ad budgets. So we created the associates program, [which lets] any Web site link to us, and we give them a revenue share. … Those things didn't require big budgets. They required thoughtfulness and focus on the customer."

Bezos further elaborated on Amazon's frugal innovation principle in a 2009 analysts' call that sounds as if he had just finished reading Buffett's shareholder letter:

"Every time we go through our planning process, we do set top-down targets for variable and fixed productivity, and we try to work towards those to make sure that we're being frugal and efficient," wrote Bezos, "so that we can have a cost structure that supports the customer experience that we want to have, which is to have the lowest prices. …"

Frugal efficiency is not ordinarily celebrated as an innovation driver; it's typically viewed as a constraint. Yet even in design-intensive, technology-rich industries, the evidence persuasively argues that investment quantity doesn't determine innovation quality. More money, more time, and more resources do not mean more—or better—innovation. Successful innovation outcomes don't depend on great wealth. Focus matters more. This concept is pure Buffett.

Focusing on fundamental value empowers innovation efficiencies. Innovators like Bezos and Jobs have remarkably clear fundamental understandings of the innovation value they seek to create. So did Tesco and Walmart. The clearer those understandings, the better focused their innovation spend is.

Bezos and Jobs are fundamental value innovators in the same way as Warren Buffett is a fundamental value investor. So are Walmart and Tesco. Their command of the innovation fundamentals of their business mirrors Buffett's command of the financial fundamentals of his. They've effectively aligned innovation investment with financial outcomes.

"In the very earliest days [I'm taking you back to 1995], when we started posting customer reviews," Bezos told the *Harvard Business Review* in 2007, "a customer might trash a book and the publisher wouldn't like it. I would get letters from publishers saying, 'Why do you allow negative reviews on your website? Why don't you just show the positive reviews?' One letter in particular said, 'Maybe you don't understand your business. You make money when you sell things.' But I thought to myself, *We don't make money when we sell things; we make money when we help customers make purchase decisions.*" [italics added]

Exactly. Bezos describes a fundamental organizing and animating innovation principle of Amazon's business model. The importance of recommendation engines, Amazon Prime, and artfully bundled online promotions is embedded in that 2007 quote. "Selling things" requires different value assessments than "helping customers make purchase decisions." Innovations that help customers make purchase decisions require different investments—and experiments—than innovations that sell things.

Successful innovation practices highlight those distinctions. Understanding, embracing, and communicating those distinctions define Amazon's frugal innovation investment culture. And frugal actions speak louder than words.

Consequently, cost reduction doesn't drive fundamental value innovation investment—fundamental value investment drives cost

reduction. Successful innovators—such as Bezos, Jobs, and Sir Ter-
ry Leahy's Tesco—aren't traditional cost cutters as much as investors seeking underappreciated innovation opportunity. These value
innovators invest with margins of safety firmly in mind.

Success comes not from making bolder or braver innovation
bets than rivals, but by carving out broader margins of safety. This
may seem counterintuitive, but it's pure Buffett. Value innovators
would rather be certain of a good result than hopeful of a great
one. Of course, good results can rapidly grow into great ones.

THE BUFFETIZATION OF EXPERIMENTATION

What cultural, organizational, and operational practices create
those margins of safety for innovation? Bezos put it best: "The key,
really, is reducing the cost of the experiments."

Cheap experimentation is the "secret sauce" of value innovation
investment. Bezos and Amazon are champions of cheap experimentation, which in turn has fueled innovative growth consistently
and successfully. Amazon is managed not just as a virtual store,
but as a global digital laboratory. Integrating experimentation and
merchandising gives Bezos the benefits of a "two for the price of
one" value proposition.

Practically every day, Amazon runs hundreds—perhaps thousands—of experiments exploring different aspects of helping customers make purchase decisions. Many are simple or incremental
tweaks in recommendations or page display; others test new genres
of interaction. Continuous experimentation is baked into Amazon's
behavior. Experiments are ongoing processes, not discrete events.
This process discipline mirrors Buffett's investment discipline.

I once asked Bezos at a CEO conference what he thought the
most important aspect of his company's experiment culture was.
He looked at me for a moment, wide-eyed, and responded, "It's
what we do. It's what we've always done. We're always doing
experiments."

But this is not experimentation for the sake of experimentation. These experiments are business investments aligned with

measurable returns. They test business hypotheses about improving operational efficiencies and the customer purchase experience. Frugal experimentation drives Amazon's frugal innovation. It makes the economics of everyday experimentation feasible as well as valuable—with large margins of safety to boot.

Left unspoken (but a core competence that Amazon's competitors understand and acknowledge) are speed and immediacy. Being cheap is not enough. Time and tempo also define value innovation. Fast is as important as frugal. If it isn't fast, it isn't frugal. The velocity of experimentation matters as much as its cost. Cheap experiments that take months to finish are typically far less valuable than those that take minutes.

SCOTT COOK, FOUNDER AND CHAIRMAN OF INTUIT, ON EXPERIMENTATION

Our head of global asked our teams in India to figure out how can we improve the financial lives of people. One team had the idea of focusing on farmers. They're half the population in India. We'd never done anything with farmers. Their bosses were uninterested. So the team went out, researched, spent time in the fields, time with the farmers, time in their homes, came up with a problem that they thought they could solve. These guys don't know where to take their goods when they harvest them. Do they take them to the town to the north? The south? The east? Which wholesaler will give them the best price? ... So our little team said, "Why don't we just collect the prices and send them on SMS so they get them on their phones?"

Well the bosses had all the reasons it wouldn't work. You know, farmers are often illiterate. Could they read it? Would they believe it? Would it change behavior? ... And would we even be able to get the price information from the wholesalers? ...

Because of [our] culture of experimentation, the team didn't listen to the bosses. Instead they ran an experiment. [In] seven weeks, they [prototyped] a product, went out and tried it. Two weeks later, they had proof it would work. One hundred and ten farmers, on average, reported 16 percent higher incomes. Twelve of the 14 wholesalers said they'd continue to give us data. Thirteen experiments later, we now have over 400,000 farmers who get price information on their phones in India. ...

The bosses would have killed it. But because of experimentation—cheap experiments—that was an idea that could prove itself. *This is a new kind of management, where instead of viewing the boss's role as the Caesar to make decisions, the boss's role is to put in a system whereby junior people can run fast and cheap experiments, so that the ideas can prove themselves.* [emphasis added]

Source: 2011 Techonomy Conference in Tucson, http://techonomy. com/2012/08/scott-cook-on-scientific-experimentation-in-business (with kind permission of Techonomy Media).

Fundamental innovators need to go beyond Buffett. "How do you buy a dollar's worth of innovation for fifty cents or less *faster*?" becomes the value innovation investment question. That's why innovation experimentation is a craft, not a science. Innovation investors must decide the appropriate trade-offs between fundamental value, speed, and margins of safety. They need to make themselves comfortable with how cheap *and* quick experiments manage innovation risk.

The young Steve Jobs provides a compelling example of how smart innovators extract value from simple, fast, and cheap experiments. In Apple's earliest days—long before the iPhone or iPad was a gleam in his discriminating eye—Jobs identified key technologies to commercialize and popularize. He was brilliant at this.

The Alto computer—conceived in Xerox's remarkable Palo Alto Research Center back in the late 1970s—embodied the personal computing aesthetic that Jobs desired. He was particularly intrigued by the handheld mouse as an interface device. The mouse would be his first user-friendly breakthrough interface for the masses.

But there was an expensive problem. The cost of the goods that went into making the 1980s Xerox mouse topped $400, and Jobs's vision required a cheap, mass-produced mouse. So he turned to a small Stanford University industrial design/mechanical engineering spin-off called Hovey-Kelley. (This small team eventually became IDEO, one of the world's most successful industrial design firms.)

Jobs's design instruction to Hovey-Kelley was demanding: create a Xerox mouse equivalent whose cost of goods would total less than $25. In other words, Jobs wanted over 80 percent of the Xerox mouse feel, form factor, and functionality for less than 10 percent of its physical cost. This was a Buffettian fundamental value challenge of the first order.

Hovey-Kelley quickly—and cheaply—rose to the challenge. "The first place I went was to Walgreen's," recalled Dean Hovey, one of Hovey-Kelley's partners. "I bought all the roll-on deodorants I could find on the shelves. They had these plastic balls in them that roll around. ... I hacked together a simple spatial prototype of what this thing might be, with Teflon and a ball. The first mouse had a 'Ban Roll-On' ball."

The prototyped packaging of these off-the-shelf components validated the key business hypothesis. The mouse could be mass-manufactured within fine tolerances at convenience store prices. This prototype quickly became the design template for a cheap, reliable mouse that ultimately incorporated optical encoder wheels, a free-moving trackball, and an injection-molded frame. Hovey-Kelley helped make Apple's mouse a commercial and technical success for Jobs.

The circumstances surrounding experimental success deserve a deeper look. The journey from design brief to prototype presentation took less than a week, according to Hovey. Jobs was sophisticated enough to recognize that this plastic improvisation addressed his mass-manufacturing concern. The prototype persuaded him that Apple could make good money safely by creating a knock-off of the Xerox mouse. He hired Hovey-Kelley to finish the job that its experimental prototype successfully started.

Let's scratchpad the innovation economics of this value experiment. The numbers are rough but revealing. Jobs's approach sharply contrasts with how more-established technology firms—IBM, Xerox, and AT&T, for example—might have pursued the task of making mice profitably.

Hovey-Kelley received its design brief and crafted a prototype that confronted its client's core business concern in five days. That prototype didn't solve the problem. But it demonstrated that the problem could be solved with a fairly high level of confidence. The prototype earned buy-in from a difficult and discriminating client. The time and material costs for this rapid experiment were less than $1,000.

An IBM or Xerox innovation scenario would have played out differently. The design brief likely would have been a detailed request for proposal (RFP) which would invite design and contract manufacturing firms to present plans and budgets explaining how they'd mass-produce mice.

The client wanted vendors to solve the $25 mouse problem in a provable fashion. Of course, the contractors had never seen, let alone built, such a device. They'd perform cost of goods analyses and compare manufacturing options.

This procurement process—whether internally managed or outsourced—takes serious time and money. RFP preparation requires at least a couple of weeks. So does the task of circulating and responding to questions. Processing, reviewing, and comparing the rival RFPs typically takes another month. Having final candidates present their proposals adds another fortnight after that.

In sum, procurement and due diligence time investments to seriously launch mouse development require roughly 100 days. Hard- and soft-dollar costs—even in 1980s dollars—would approach $100,000.

This "big-company" analytical approach is arguably more thorough and comprehensive. Let's say that the client enjoys a probable 90 percent confidence level in mass-producing that $25 mouse by heavily investing in RFP due diligence. The faster, cheaper, but far less comprehensive "hypothetical prototype/prototyped hypothesis" approach merits, by contrast, perhaps a 75–80 percent confidence level.

The bottom line: Big Company spends $99,000 more money and 95 days more time, to gain maybe 15 percent more confidence in its ability to produce mice. That's a significant gap. But is it a good investment? Is it worth it? The most honest answer is: "We don't know." The most important technical and commercial challenges may lurk within that 15 percent differential.

But the rapid experimentation route still leaves the aspiring mouse innovator with $99,000 and 95 days. Quick experimentation doesn't solve every problem, but it moves the organization further and faster, with resources to spare. Hovey-Kelly and Apple still would have lots of time and money left to address outstanding uncertainties.

Opportunity costs are minimized. No significant sacrifice of either time or money is required. Compared to more formal procurement processes, simple, cheap, and quick models generate actionable insights exceptionally fast. There may be a long way to go, but alternative options remain open. Time, money, and other resources have been preserved to be directed where they're needed.

Now, it's true that the informational business value of the model depends on the expertise—i.e., the circle of competence—of its designers. But that's like saying that successful investment depends on the circle of competence of fundamental investors. Indeed, that's what Buffet says.

Comparably, it's true that not every client can appreciate the fundamental value of the prototype. Steve Jobs evaluates prototypes with the same acuity that Warren Buffett brings to balance sheets.

But this idea affirms Buffett's overarching theme about successful fundamental value investing: people matter. Successful investors and successful innovators have a lot in common. They are committed to cultivating a discipline and expertise around fundamental value. Recognizing and investing in undervalued experiments as an asset class are actions that smart fundamental value innovators take. They see how an underpriced model can yield information insights that are worth a lot of money.

The Ban Roll-On mouse is, literally and figuratively, a model fundamental value asset. Its margins of safety were terrific; its return on investment enviable. Apple—and Hovey-Kelley—reaped well over a dollar's worth of innovation for every fifty cents they invested.

More accurately, this rapid prototyping experiment purchased more than a dollar's worth of innovation information for far *less* than fifty cents. This distinction is important. Innovation information and insight constitute the coin of this experimental realm. In Apple's internal investment market, Hovey-Kelly's rapid experiments helped strip roughly 90 percent of the cost from the $400 Xerox product. A simple innovation investment delivered economic value far faster and cheaper than other alternatives.

From a Buffettian perspective, this is fundamental value engineering as fundamental value investment. The investment experiments respected and reflected circles of competence and margins of safety. Apple managed innovation risk and reaped innovation rewards profitably by honoring fundamental value investment principles.

Established firms are as able to exploit the favorable fundamental economics of experimentation as can entrepreneurs. Procter & Gamble (P&G), for example, has culturally, organizationally, and operationally redefined its innovation economics of concept testing.

For decades, Peoria, Illinois—a self-described All-American city—was the nation's capital of new product experiment and exploration. "Will it play in Peoria?" was the question that dominated marketing debates by ambitious brand managers anxiously promoting new ideas. Companies invested inordinate resources toward maximizing Peoria's value as a test marketplace. P&G, always a new product pioneer, practically set up a subsidiary there.

The real Peoria has become a marketing anachronism, though. The Internet has become a virtual Peoria for P&G. The world's largest consumer packaged goods company has been making the Internet a digital laboratory and test marketplace for value innovation.

"We do almost 100 percent of our concept testing online now, at literally one-hundredth of the cost and one-hundredth of the time," asserted Steven David, a P&G chief information officer (CIO), back in 2003.

Even discounting for hyperbole, P&G's ability to concept-test quickly and cheaply transformed its economics of innovation. The company can do much more for much less, and much faster. Win. Win. Win.

In theory, P&G could explore 10,000 times more concepts meaningfully today than it could a decade ago. In practice, digital experimentation and test creates huge margins of safety for exploring not just new product ideas, but new product categories. That revolutionizes innovation expectations inside the firm. The speed and cost of iterative experimentation—refining, tweaking, and focusing on fundamental innovation attributes—energizes creativity.

"We used to do a lot of tests and experiments to confirm our value analysis," said one longtime P&G marketing executive. "We've become much more comfortable testing in ways to learn things that can also surprise us. We can get more kinds of value in much less time."

The investment success of cheap, simple, and fast experimentation flips an important fundamental principle of modern portfolio theory on its head. The conventional finance argument insists that great returns can be obtained only by taking great risk. This is rubbish. Fundamental innovators at Apple, Amazon, and P&G know that the opposite is true: innovators who can identify experiments with frugal and fundamental value can reap high return with relatively low risk.

This principle is enormously important: Innovators are capable of developing great value without paying exorbitant premiums or taking dangerous risks. Investing in underpriced or undervalued experiments is much like investing in an undervalued or underpriced stock. The upside potential is remarkable; the downside risks are marginal. The quest to purchase a dollar's worth of innovation information for less than fifty cents isn't fanciful; it's pragmatic.

Fundamental value innovation may fly in the face of the 'efficient market hypothesis' beloved by finance departments and business schools. But the reality is that most organizations have an embarrassment of potential experimental riches. There's absolutely no shortage of "fifty-cent" innovation dollars. But there is a shortage of individuals and teams who have the wit and willingness to identify and invest in them. The next chapters explore how this is done.

6 INVESTING IN EXPERIMENTS

Experimentation is the least arrogant method of gaining
knowledge. The experimenter humbly asks a question
of nature.

—Isaac Asimov

Galileo's Tower of Pisa. Newton's prisms. Mendel's peas. Foucault's
pendulums. Pasteur's flasks. Wilson's cloud chambers. History
suggests remarkably simple, cheap experiments can profoundly
transform the sciences. *The Innovator's Hypothesis* proposes that
remarkably simple and cheap experiments can similarly transform
industries and start-ups. Curiosity and ingenuity matter more than
budgets.

Of course, science isn't business, and business surely isn't a science. But the potential and power of a single experiment to radically disrupt existing perceptions and expectations is common to
both. That's as true for physics and molecular biology as financial
services, retail, chemical engineering, or biotechnology.

What's more, the raw materials and technical ingredients needed for effective experimentation have never been more accessible,
affordable, or scalable. Pragmatically, these new economics of
experimentation are fantastic. Scientists in a multitude of disciplines can run more experiments, by orders of magnitude, at lower

osts, again by orders of magnitude, faster than ever. New technologies, for example, have reduced the costs and time of DNA sequencing from thousands of dollars and days to "the time and reagent cost of sample preparation," as one study put it (http://www.ncbi.nlm.nih.gov/pmc/articles/PMC3337438). Has there ever been a better time to explore a testably disruptive—or disruptively testable—business hypothesis? The opportunities for enterprises and entrepreneurs alike to turn portfolios of simple experiments into transformative, value-added, and cost-effective innovations are there to be seized.

"The reason why we are on a higher imaginative level is not because we have finer imaginations, but because we have better instruments," declared Alfred North Whitehead, the great English mathematician and philosopher. "In science, the most important thing that has happened over the last forty years is the advance in instrumental design. A fresh instrument serves the same purpose as foreign travel, it shows things in unusual combinations. ... The gain is more than a mere addition; it is a transformation."

A transformation. What's so striking about Whitehead's comment is its 1925 provenance. Back then, transistors, microprocessors, and Internets weren't even glimmers in the eyes of science fiction writers. But the brilliant philosopher and mathematician clearly grasped the fact that technical capability—not just intellectual brilliance—was essential to enabling insight.

Ironically, Whitehead's observation proves more valuable as pre-science than historical review. Today, a mass proliferation of tools and technologies is digitally democratizing even the most sophisticated scientific instrumentation. "Maker-spaces," with 3-D printers, bio-hackers, and cheap apps that turn tablets and smartphones into sophisticated sensors and scanners, essentially mainstream what was once elite science. The opportunities for amateur and semi-professional innovation have expanded exponentially.

"Many signals exist today that point to a potential shift in the next several years of where scientific discoveries come from and who can actively contribute to science," writes Ariel Waldman, "science hacker" and author of *Democratized Science Instrumentation*, a 2012 white paper. "[This movement] is less about replicating lab equipment and more about being empowered to create new instrumentation to explore the often overlooked, underfunded, and fringe areas of science" and technology.

Does anyone believe that the next decade of instrumental innovation in science and technology will be less interesting or provocative than the last? On the contrary—Whitehead's "advance in instrumental design" ensures a perpetual revolution in which innovators' hypotheses can thrive. Tiny but targeted investment in simple, fast, and frugal experiments will reshape how organizations create new value for themselves and their customers.

Science and scientists, however, take experimenting far more seriously than business and businesspeople. The burdens of business history, culture, education, training, technology, and economic inertia deemphasize experimentation in the hearts, minds, and budgets in the enterprise. Even the simplest and cheapest experiments face a reflexive organizational skepticism and resistance. There are no CEOs—not if we mean "chief experiment officers."

"The real measure of success is the number of experiments that can be crowded into 24 hours," superinventor Thomas Alva Edison once insisted. Alas, the Wizard of Menlo Park's innovation metric has the musty aroma of industrial nostalgia rather than the cool breeziness of best business practice. In truth, business experimentation remains undervalued, underappreciated, and underexplored, both as an asset and an investment opportunity. Why? Because most organizations don't know how to either design or manage portfolios of business experiments. They don't grasp the new economic fundamentals of experimentation. They lack the confidence, competence, and culture.

Misunderstanding and mispricing experimentation's potential value is wildly expensive and counterproductive. It's inefficient; a waste of time, talent, and money. It undermines opportunity and hurts morale. It's 'buying high and selling low.' The inability and reluctance to perform even simple experiments is not unlike an inability or reluctance to respond respectfully to customer complaints. It's fundamentally toxic.

Turning managers into Galileos, Newtons, or Edisons isn't the answer. That's both ridiculously unfair and unrealistic. Executives aren't scientists—and they shouldn't be. The smarter and more serious challenge asks businesspeople to revisit the fundamentals: that is, to become better investors. More specifically, organizations need to become smarter, more serious investors in fast, frugal, and simple experiments that test high-impact business hypotheses. Experiments need to be seen and managed as investments.

As superinvestor Warren Buffett once noted, "There is really a lot of overlap between managing and investing. Being a manager has made me a better investor, and being an investor has made me a better manager."

This book puts a behavioral snap in Buffett's value tale. Replace the word *managing* with *experimenting*. That simple substitution alters organizational, cultural, and operational expectations around innovation investment. When you treat experiments more as investment assets than research exercises, you value them differently: "There really is a lot of overlap between experimenting and investing. Being an experimenter has made me a better investor and being an investor has made me a better experimenter."

In other words, if your proposed experiment were a business, how good an investment would it be? How do you know? Would you be thrilled, happy, or merely satisfied with its potential payout? What are the chances—the risk—the investment proves utterly worthless? Every member of every 5×5 team needs to be able to answer these questions. They're the essence of fundamental value innovation and the fundamental value proposition of *The Innovator's Hypothesis*.

Science experiments seek insight into fundamental truths; business experiments seek insights into fundamental value creation. People need to appreciate and understand the fundamentals. The right experiments make that awareness possible. Facilitating greater insight into business fundamentals typically generates superior returns on investment.

Think of the 5×5 experimentation methodology as a mash-up of financial theory, design thinking, and the scientific method. Financial theory—inspired by Nobel laureate economists such as Harry Markowitz, Bill Sharpe, Eugene Fama, Daniel Kahneman, and Vernon Smith—yields technical awareness of the risk/reward ratios, heuristics, and transaction costs governing investments in experiment portfolios. Finance offers a powerful conceptual framework and metaphor for evaluating the mix and management of experiments.

Where modern portfolio theory posits that investors can construct equity portfolios that maximize possible returns relative to perceived risks, the 5×5 proposes that innovators can craft experiment portfolios that maximize actionable innovation insights for managing and mitigating innovation risks. Innovators and investors alike need to think carefully about what diversifications best balance their market ambitions and risk appetites.

No, the financial metaphor isn't perfect or exact. But it effectively disciplines portfolio design discussions in the business contexts of risk and reward. That's important. Managements should be smarter investors in simple experiments. The essential principles of financial theory apply directly to experimental design.

Design thinking frames the most desirable insights and desired outcomes from experimental designs. From Nobel laureate Herb Simon's book *Sciences of the Artificial* to IDEO CEO/president Tim Brown's "Design Thinking" philosophy design becomes a way to make experiments more accessible and appealing inside the enterprise and out. A design sensibility helps ensure that neither individual experiments nor experiment portfolios become too narrow or reductionist. They're explicitly designed with broader UX (user experience) and organizational impact in mind.

For one California firm, design thinking shifted the focus of 5×5 teams from solving problems toward enhancing experiences. The automobile advisory company's portfolio investment theme centered on engagement. How could the organization rethink, redefine, and reshape how it engaged with its customers as they shopped for cars?

Focus groups were out; so were detailed questionnaires. Design thinking required understanding how their customers behaviorally drew distinctions between communication, interaction, and engagement. The 5×5 teams designed experiments exploring engagement around giving advice, not just providing information.

Design thinking, for one team, meant experimenting with how the company could use instant messaging (IM) to advise car buyers. Should IM advice be structured or improvised? Should the advisor be a just-in-time resource or a real-time partner? How should IM support be branded as an experience? Could IM advisors and IM advisories be done in ways that wouldn't offend the auto dealers?

IM offered a simple, cheap, and accessible platform for real-time engagement experiments as customers shopped. By making "engagement around advice" their investment theme rather than "customer support," the team came up with a portfolio of experiments that successfully provoked broader conversations around the firm's brand promise to customers. Marrying design thinking to experimental design improved the quality of both.

As the great graphic designer Paul Rand observed, "To design is much more than simply to assemble, to order or even to edit; it is to add value and meaning, to illuminate to simplify, to clarify, to modify, to dignify, to dramatize, to persuade and perhaps even to amuse. To design is to transform prose into poetry."

This book champions that experimental design ethos. Empathy and creativity are as integral to crafting experiments and testing hypotheses as rigor and rationality. The poetry of well-designed experiment portfolios can inspire remarkable innovation outcomes.

Even better, thoughtful design investment dramatically improves the odds for serendipity. To paraphrase Pasteur, chance favors the prepared portfolio.

The scientific method, of course, enables rigorous, reliable, and replicable construction of business hypotheses for experimentation and testing. The *Oxford English Dictionary* characterizes it as "consisting in systematic observation, measurement, and experiment and the formulation, testing, and modification of hypotheses."

But French physiologist Claude Bernard's formulation fluently captures experimentation's appeal as organizing principle for insight: "Observation," he noted, "is a passive science; experimentation is an active science."

The 5×5's entire thrust makes experimentation the interactive integrator of financial theory, design thinking, and the scientific method for innovation. But there's a crucial methodological difference between 5×5 experimental design and what serious scientists do. That critical distinction is economic constraints.

Look at elite science journals like *Nature, Cell,* or the *Proceedings of the National Academy of Sciences*. Read first-person papers by world-class scientists describing their Nobel-winning experiments: for instance, Watson and Crick's groundbreaking "Molecular Structure of Nucleic Acids: A Structure for Deoxyribose Nucleic Acid," or "Production of Coherent Radiation by Atoms and Molecules" by Charles Townes, inventor of the laser. Glance at Newton's *Principia* or publications by James Clerk Maxwell, Ernest Rutherford, Albert Abraham Michelson, Kary Mullis, Enrico Fermi, Linus Pauling, Joshua Lederberg, Frederick Sanger, and other pioneers in their fields. You'll quickly notice that something's missing. Something important.

Virtually none of those narratives detail in any meaningful way the budgets and costs that went into their experiments. There's virtually no discussion about the economics of their breakthrough experiments. It's as if money doesn't exist. As a rule, custom, and proud culture, scientists don't submit their spreadsheets or schedules for prepublication peer review. Similarly, peer reviewers rarely look at papers with an eye toward suggesting radically faster or cheaper ways to experiment. Economic and logistical concerns and challenges are incidental, not central, to what's formally published as science.

As one MIT molecular biologist confided, "Of course I care about the cost of experiments; they can be time consuming and expensive. ... But they're not relevant to what we publish. ... When I'm thinking about patents or commercialization, then I think [more seriously] about how we might do some of our experiments differently." This view is not atypical.

The 5×5 approach creates different real-world contexts for harvesting economic value from the scientific method. Where serious scientists chafe at scheduling and economic constraints, the 5×5 makes a virtue of necessity: time and budget limits require collaborative ingenuity around fast and cheap experimental design. The challenge isn't crafting the best possible experiment to test a business hypothesis—it's crafting the best simple experiments with a minimum of time and money.

In software development parlance, 5×5s turn the bug of seemingly arbitrary constraints into the features of simple, cheap, and fast experiment portfolios. This links directly to design thinking.

Design depends largely on constraints. Here is one of the effective keys to the design problem: the ability of the designer to recognize as many of the constraints as possible (and) his or her willingness and enthusiasm for working within these constraints. By insisting on constraints, 5×5 portfolios dramatically improve the likelihood that they will generate high-impact, high-value returns on investment.

Consequently, blending these three sensibilities represents neither a kludge nor shotgun marriage of incompatible perspectives. To the contrary, the combination of finance, design, and science facilitates a virtuous cycle of innovation information and insight. But it's imperative, therefore, to appreciate that this book is not simply (or even primarily) about doing or designing experiments. In fact, this is more a book about how to invest in experimentation; tt's an investment guide. Organizations shouldn't just ask what kind of experiments they should be doing. They need to answer the question: What kind of investors do we want to be?

Being a good investor in experimentation is not the same as being a good manager of experiments—and vice versa. But one needs to address both aspects, often at the same time.

What does the organization need to learn fast? What does the organization want to learn fast? How does the enterprise really see and define risk? What risks are you prepared to take to understand risk better? What business hypotheses inspire the most excitement? Which ones provoke the greatest anxiety?

Crudely put, if the experiment were a stock, what kind of stock would it be? A blue chip? Momentum? Value? Growth? What does your experiment portfolio reveal about how you value opportunity, risk, and diversification? How safe or risk-averse is your portfolio? Looking at it another way, do you see your experiment portfolio as your opportunity to take novel risks to build a new innovation future?

Some organizations will effectively "buy the index" in a "me, too" effort to make sure that they remain competitive. "Good enough" experimentation is good enough for them. Other firms will feel competitively compelled to outperform their rivals and embrace riskier portfolios. Still others will look to their customers and clients and design experiments to test new value and new markets in creative ways. You can be sure that their business hypotheses will look nothing like those of the indexers.

The challenge comes from having the courage to confront these questions honestly. The opportunity comes from recognizing there is no better medium for managing innovation risk and exploring new value options than investing in smart, simple experiment portfolios.

But in the final analysis, investing in experiments—like investing in stocks—is a bet on the future. As Warren Buffett repeatedly tells investors, "Price is what you pay; value is what you get." Fusing financial theory, design thinking, and constraint-based science, *The Innovator's Hypothesis* is an investment guide for organizations intent on getting significantly greater value for less money, time, and risk. It's a good bet.

THE
INNOVATOR'S
PORTFOLIO

7 BLOCKBUSTED: A CASE STUDY IN EXPERIMENTAL FRUSTRATION AND FAILURE

Nullius in verba
"Take no one's word for it."

—The motto of the Royal Society

It didn't make sense. Why would a smart company—a market leader—spend millions on a legally risky and controversial solution to its most serious business and branding problem without even attempting a simpler and safer $10,000 experiment?

Beats me. But that's the bizarre position I found when I started working with Blockbuster in late 1999 to 2000. After years of complaints, the multibillion-dollar video rental giant finally confronted its most painful customer service issue. I was invited to lend a hand.

Blockbuster's challenge was clear. Survey after market survey confirmed that—more than anything else—the company's customers hated late fees. (Blockbuster called them "extended viewing" fees; that euphemism didn't soften the blow.) Nothing infuriated customers more—not long lines, stock-outs, or surly unhelpful staff. People resented having to pay extra for their own forgetfulness or poor planning. The more frequent their fees—fines, really—were, the more furious they got. They didn't blame themselves; they blamed

Blockbuster. So Blockbuster had lots of angry customers. One of those customers, Reed Hastings, was so outraged, he became an entrepreneurial avenger. After being charged $40 for being late in returning *Apollo 13*, Hastings went on to create Netflix.

The data I saw also indicated that Blockbuster's biggest offenders were often its best customers—the people who rented the most films the most often. These premier renters didn't suffer in silence. Market research suggested they badmouthed Blockbuster to friends and acquaintances (even as they continued renting). Store managers regularly reported furious customers yelling at employees about these fees. Blockbuster was even sued over its late-fee policies. I'd never seen a customer service situation like it.

But this seething "renters' rebellion" coexisted with an irrefutable business fact: the money was great. Late fees were remarkably profitable. Blockbuster never shared exact numbers, but analysts at the time estimated that over 20 percent of the company's pretax profit came from extended viewing fees. Blockbuster literally made tens of millions in profit thanks to customer laziness, forgetfulness, and oversight. The firm did nothing and got generously compensated. A sweet deal if you can get it.

Easy money inspired insidious innovation. Blockbuster management seemingly became addicted to "two-for-the-price-of-one" and "three-for-the-price-of-two" promotions. Entice more people to rent more movies at a time. The bet was that a nontrivial number of these additional "free" rentals would be returned late. That bet paid off handsomely; those promotions surged revenues from neglectful customers.

This wasn't a secret. Everyone at Blockbuster knew it, and so did its customers—and its competitors. An entrepreneurial "video rental by mail" start-up called Netflix was taking dead aim at Blockbuster. It promised, "No late fees." The reckoning had arrived. The easiest money Blockbuster made had become a difficult problem.

Management acknowledged its status quo was unsustainable. Between lawsuits, customer dissatisfaction, and intensifying competition, they agreed something had to be done. People both inside the company and out were told to come up with solutions to the "extended viewing fee" headache.

The brief was straightforward: Help Blockbuster transform late fees from a primary pain point into a marginal concern for the company and its customers. Everything was on the table, they claimed. Be innovative. Be creative. Challenge us. We know we have to change. (But don't make things too complicated for our stores to implement.)

As context, Blockbuster told me of several bold and transformative proposals for late fee management. By my criteria, these seemed ambitious, complicated, and expensive. For example, have customers buy stamped, preaddressed envelopes so they could mail back overdue movies; or allow over-the-phone renewals for a fee. These initiatives seemed interesting, but each of them would be costly to test. I thought we could do better.

I reviewed Blockbuster's data, spending hours in its stores. I listened to customers and clerks alike talk about extended viewing fee behaviors and misbehaviors. I didn't do extensive surveys—that wasn't my assignment. Rather, I paid close attention to the everyday Blockbuster operations. Based largely on the 80/20/20 principles that would evolve into the 5×5 framework, my suggestion was anything but comprehensive. It had the virtue of strategic specificity.

My modest proposal: a simple, scalable experiment with potentially big impact. Fast, cheap, and easy was my design philosophy. Based on what I'd studied and observed, Blockbuster needed to learn much more about customer behaviors—and fast.

The company simply didn't know enough about its habitual "late-fee" customers to launch sweeping, big-budget solutions. That was risky and presumptuous. A focused customer-centric experiment was the better bet. Running a quick but careful experiment would buy Blockbuster time and insight on its path to more comprehensive solutions. Better swift and safe than expensive and sorry.

My experiment explored what appeared to be an obvious yet unasked question: Did Blockbuster's customers want to be reminded to return their movies on time? Were they willing to help Blockbuster help them avoid extended viewing fees?

The answer was a question: Why not ask? Why not respectfully involve them? Good-faith efforts to help Blockbuster customers avoid late fees might make the company more likable—or at least customers might dislike it a little less. Either outcome would be a win. More important, Blockbuster would gain valuable insight into its customer relationships.

We'd use the fast-growing medium of email as our Customer Reminder Technology (CRT) laboratory. Email was cheap, pervasive, and fast-becoming a bigger part of customers' lives. Blockbuster was already exploring email marketing. (That was, I thought, good news for my proposal's implementability.)

This concept cut to the customer engagement core of Blockbuster's unhappiest moment of truth. Might email nudges actually work? The tentative hypothesis was that more customers indeed would return their movies on time if reminded to do so. But how many? Which ones? What perceived—and real—difference would reminders make?

Perhaps this service could help rebrand Blockbuster as friendlier and more customer-sensitive. Maybe tardy customers who'd received email reminders would feel more annoyed with themselves than with Blockbuster for late returns. Reducing in-store arguments over late fees might boost employee morale. Surely Blockbuster would care about these issues.

From my perspective, this was a no-lose proposition for the client. Reminders—for presentation purposes, I called them "Eminders"—couldn't help but give Blockbuster actionable insights into its most controversial profit center.

Structuring the experiment was straightforward. Design an email reminder protocol. Pick a representative sample of ten or twelve Blockbuster stores. Have store staff ask customers for their email addresses as they rented movies at checkout for two or three consecutive weekends. Write a script that they would use to explain that addresses would be used solely for sending reminders the night before the films were due back—no promotions, no solic-itations, no spam. See how many people signed up. Give those addresses to a central information technology (IT) or customer care team responsible for sending out Eminders and tracking their effectiveness. Compare return rates and late fees against a control sample of comparable stores. (Remember that email use in 2000–2001 was rising fast, but not yet ubiquitous. Blockbuster hadn't made customer email capture part of its regular business practices. The Internet was a nascent part of Blockbuster's business model at that time.)

This experiment would be cheap. A no-frills version would require less than a $12,000 investment in hard dollars. My back-of-the-envelope calculations suggested that this would represent barely a tenth the cost of testing rival late-fee options. The meth-odology and math were simple. No special expertise, training, or technology was necessary. Blockbuster could run it in house or outsource. Perhaps only a single networked PC would be required.

Eminders also set the stage for exploring future email and Inter-net customer relationship management possibilities. Store manag-ers and staff could be encouraged to suggest further refinements and tests. Cheap experiments could add value to existing Block-buster data. I (rather immodestly) believed that this would be a strategic bargain at five times the price (including my fee).

I thought this proposal delivered everything that Blockbuster's brief had asked for and more. The likely costs and complexities of rival extended viewing options increased my confidence that Blockbuster would find this proposal great value for money. At the very least, I expected that Eminders would be received with cau-tious enthusiasm. I could not have been more wrong.

The people in that conference room looked at me as if I had peed on their carpet. Their eyes narrowed. They were neither happy nor impressed. After a few moments of irritated silence, the Blockbuster attendees spoke up. Not one comment addressed any cost or technical design issues. No one suggested enhancements or improvements to the core concept. My proposal may have satisfied the brief, but it didn't satisfy this audience. They hated it; they hated *me*.

Their questions were hostile and accusatory: *Why on earth do we want to remind customers the day before to return their movies on time? Why should we test an idea that's virtually guaranteed to reduce our profitability? How will this help us make up for the money we'd lose?*

I was stunned. My immediate reaction was defensive. This was a simple experiment, not a grand strategy. We weren't prototyping a national Blockbuster email reminder system. We were testing a simple technical innovation to gain both business and customer insight fast. Focus on the goal: learn important things about rental return behavior as quickly and cheaply as possible. Whatever path Blockbuster ultimately chose, experiments like this would help reduce implementation risk.

Besides, how harmful could such an experiment be? Perhaps customers would rent even more movies if they believed that Eminders would lower late fee risk. Simple experiments could help Blockbuster redefine customer satisfaction around its most contentious issue. Focus groups alone never could get you there. Actions speak louder than words.

I talked faster: Exploring email connectivity with customers would help stores. We could sharpen the company's web strategy as competition with Netflix—that "No late fees" upstart—intensified. Maybe Blockbuster could use this experiment to launch online renewal testing. Put aside your obvious annoyance, I practically pleaded, and consider both the brief and the numbers: Blockbuster won't find a simpler or easier way to learn more about its customers faster.

Nothing. Sullen silence. This group didn't care. They had zero interest in any hypothesis that my experiment could test or any insight it might generate. The harder I sold, the more irritated they became.

I went down in flames. The people in that room didn't want to observe customer behavior at all, whether in store or via email.

They felt no need to run in-store experiments. To my unpleasant surprise, cost was irrelevant. I could have offered to pay for the entire experiment myself, and they still would have rejected it. They wanted this issue to disappear. Customer unhappiness was a nuisance to eliminate, not a business issue to understand. They wanted solutions that placated or pacified angry customers. But they really liked those late-fee revenues.

My cheap little proposal exposed the brief's naked irrelevance. To my outsider's mind, these managers hadn't thought through what they wanted to learn or needed to accomplish. They didn't define business processes or operational trade-offs around deemphasizing late fees. Every single thing that this experiment might explore, they vehemently resisted. They didn't want to know. Wow.

I was so screwed. None of the other bold or transformative proposals under review, of course, had even a speck of supporting data or analysis. They were grandiose "Big Idea/Big Budget" plans that demanded big up-front investment. The cost of analyzing and testing them would zoom into the hundreds of thousands of dollars, and the testing would take many months. These initiatives were the antithesis of simple, cheap, and fast. Blockbuster would pay lots of money and take lots of chances testing proposals with questionable odds for success.

I was beaten. The executive who'd invited me remarked later that his colleagues had wanted bigger thinking. They didn't like what they saw as my incremental or piecemeal approach to their strategic challenge. They wanted final solutions. So I apologized for wasting everyone's time and waived my fee.

But I had to ask: Why was his group so sure that "big bang" strategic approaches were best? Both on paper and Excel, my experiment offered far greater bang for the buck—and faster—than the grand plans did. Why did a simple, cheap experiment provoke such visceral negative reactions? What hot button did it push?

His response said it all. "We think we know what the problem is," he said. "We want a plan that's big enough to solve it without putting our revenues at risk. You didn't help us with that."

What could I say? That the conversation revealed Blockbuster hadn't honestly defined either its problem or the brief from a customer perspective? That, in fact, the company wanted late fees far more than it was willing to admit? That the very idea of a customer-centric experiment on this issue evoked greater anger than interest? My view into Blockbuster's innovation culture couldn't have been clearer. Inaction speaks louder than words, too.

I left a failure. Blockbuster's people—the company's culture—proved impervious to fast, simple, and cheap experimentation. In the final analysis, my recommendations weren't worth their time or consideration.

The company ultimately went with a transformative, big-budget effort to eliminate late fees. Backed by a huge advertising campaign, the plan proved both unpopular and—ultimately—illegal. Blockbuster's customers hated it, and the legal system hated it even more. No fewer than three state attorney generals filed class-action suits against the firm. Amid blizzards of bad publicity, the company settled. Estimated legal and business costs of this "extended viewing fee" fiasco exceeded tens of millions of dollars.

The core late-fee problem rematerialized. But now the firm had a lower market share and a bigger, fatter, and uglier blot on its reputation. The episode reinforced Blockbuster's punitive "late fees" brand. By contrast, Netflix, Amazon, and other competitors enjoyed accelerated growth and better customer service word of mouth. "No late fees" won more converts every day.

Did I enjoy a certain schadenfreude as Blockbuster embarrassed itself during this failure? Of course. I'd been told the company had done only superficial testing before launching its big initiative. No surprise there.

Did I derive any professional satisfaction when Blockbuster finally went bankrupt? No. But that didn't surprise me either.

LESSONS LEARNED

In retrospect, the Blockbuster debacle had an enormously constructive impact on my research and consulting. I'd never been inspired to learn more from outright rejection. My embarrassing real-world failure inspired three essential insights that both redefined and reshaped my study of behavioral economics in innovation:

- A simple, cheap experiment was as valuable for the purpose of rapidly diagnosing innovation culture as for actually testing a business hypothesis.

- The experiment likely would have been much better received if it had been their idea, not mine.

- "Smart" executives would rather talk about grand plans and sophisticated analyses than help perform simple experiments. Fast and cheap experimentation is seen more as marginal, not central, to their innovation strategy.

Those takeaways now seem obvious. But getting organizations to take them seriously is key to improving their innovation capabilities radically. The takeaways are rooted in the realities of how most organizations misunderstand innovation investment. Designing and implementing quick and cheap experiments deliver remarkable returns remarkably fast. Why? Experimentation literally becomes a reality check—a market mechanism—for testing the perceived and "real" value of good ideas.

Ironically, the first and most important outcome of rapid experimental design isn't a brilliant or breakthrough experiment. The immediate value comes from insight into the organization's willingness—or reluctance—to test itself, its assumptions, and its strategies. That willingness—or reluctance—speaks volumes about the innovation culture in place.

Simply observing how people respond to proposed experiments generates remarkable insight. In diagnostic terms, experiments are like organizational litmus paper, x-rays, or MRI scans in that they assess the barriers to innovation within a firm. They reveal the hidden and light up the concealed.

Reasonably well defined experiments instantly shift innovation conversation from the speculative to the practical. Airy-fairy strategic aspirations are immediately subject to real-world rigor. Experiments are invitations to take action. Who accepted or rejected those invitations—and why—proves enormously revealing. Individual and organizational reactions to these experimental invitations, I learned quickly, could be as revealing as the experiments themselves.

The trick—and it was a trick—was making cost, time, and complexity irrelevant to how organizations evaluated experimentation. Experiments had to be so simple, fast, and cheap that objections to them would not have anything to do with their logistics. Resistance would emerge from the experiment's perceived value proposition, not its mechanics or design. Deliberately "derisking" experiments made business issues, not budgets or schedules, the focus of organizational attention. This made fast, frugal, and simple experiments fantastic diagnostics for eliciting the real reasons behind innovation resistance.

But this approach unsubtly subverts the standard shibboleths and truisms of innovation culture management. Typically, innovative organizations celebrate good ideas or—better yet—breakthrough ideas. They encourage brainstorming. They invite their talented, innovative people to think out of the box, take risks, and fail fast. Innovative organizations also listen to customers and collaborate with them. Who could disagree with that approach?

The Blockbuster experimental insights invert these innovation maxims. Instead of defining innovation cultures by what they do, innovation cultures stand revealed through what they don't do. Organizations effectively have to explain why simple, fast, and cheap experiments aren't worth doing. Put more bluntly, simple, fast, and cheap experiments create a different kind of innovator's dilemma—i.e., put up or shut up.

When organizations respond to a clever, fast, and frugal experiment that tests an important business hypothesis with "We can't do that because ..." or "We won't do that because ...," whatever follows the word *because* shows the real innovation culture.

We won't do that because ...

- It costs too much.
- The lawyers won't let us.
- Our biggest supplier would be upset.
- Marketing wouldn't like it.
- That's not the business we're in.

Resistance offers a remarkably clear window into understanding the internal culture and organizational dynamics of innovation culture. It describes how organizations decline to invest in learning and risk management. What makes managers reject fast and frugal innovation insights? Are the reasons rational? Do they make sense? Or do they reflect organizational pathologies and dysfunctions that better be addressed? Most of the answers tend to be unhappy.

At one heavily regulated financial services firm, a reflexive fear of "legal" (i.e., the legal department) killed virtually every interesting experiment that the 5×5 teams proposed. Intriguingly, no one from legal had been invited to participate. At an industrial manufacturer, by contrast, research and development (R&D) proved the most indefatigable source of resistance. No MBA or rocket science degree was required to see that R&D suffered from Not Invented Here (NIH) syndrome. External innovation proposals were visceral threats, not business opportunities. Clarifying R&D's relentless 5×5 rejectionism ultimately led to leadership changes.

Of course, there's nothing inherently wrong with rejecting promises of fast and frugal experimental insights—especially if better options exist. But there's everything wrong with dysfunction-driven resistance that consistently chooses ignorance over insight. More often than not, "We can't do that because ..." has less to do with organizational capabilities than enterprise inertia, politics, self-perception, or all three.

To my surprise, my Blockbuster debacle's most enduring insight is that simple experiments create the most practical way to map organizational resistance to innovation. I've found no faster, better, or cheaper way to plumb the darker depths of innovation culture. Ironically, 5×5 experiments portfolios end up being remarkably valuable *even if they're never run!* They're mirrors that let people reflect on their refusals to even inexpensively innovate.

These "we're not going to do it" discussions are fascinating— especially when experiments address mission-critical enterprise themes. Often for the first time, organizations have frank and open conversations about what they're trying to accomplish. Tacit assumptions become explicit. Managers declare that even fast and frugal experiments aren't worth running because they "know" what the outcomes would be. Everyone comes away with a clearer sense of *why*—why people believe some investments are more valuable than others; why some people believe some departments are more capable than others; why some people believe some opportunities are riskier than others.

The conversations are almost always difficult. They ask managers and leaders to become more honest—or, at least, less dishonest—about their real reasons for resistance. Or what they anticipate—rightly or wrongly—as sources of resistance. They force organizations to see themselves from uncomfortable angles. Fast and frugal experiments portfolios effectively preempt the typical, reflexive rejections of, "That's fascinating but too expensive" or "By the time we finish analyzing the results, it'll be too late." Organizations have to come to grips with their non-economic barriers to innovation investment.

Of course, I've seen many organizations incapable or unwilling to have either serious or honest debates about why they reject fast and frugal ways to gain innovation insight. Effective diagnosis in no way guarantees a cure. Self-knowledge does not guarantee self-improvement. But quick, simple, and cheap investments in self-knowledge can deliver healthy value fast. As the Blockbuster example starkly illustrates, fundamentally dishonest discussions around strategic challenges invariably lead to unhappy outcomes.

Honesty is typically the better policy. So is transparency. The innovators and executives most open to fast and frugal testing enjoy the greatest credibility from 5×5 investment. At one California retailer with a major web presence, top management proposed a complex and comprehensive strategic repositioning. The grand planning was detailed, extensive, and expensive. But 5×5s conversations had identified innovation risks and opportunities that planning had not. Instead of ignoring or dismissing experiments as distractions, top management adopted a few as reality checks. The entire company saw that top management was open to fast and frugal learning. Aligning grand planning with simple experimentation better defines how innovation cultures change.

The most poignant aspect of 5×5 resistance mapping, however, is that organizations ultimately reject experiment portfolios they themselves design. The firm's most talented people discover—for better and worse—what their organizations simply won't do. This can be a shock. But there appears no clearer way of cultivating a culture of fast and frugal experimentation than inviting 5×5 creativity and collaboration. Innovators need to understand their culture of resistance.

OVERBOOKED AND OVERWHELMED

My favorite story illustrating those themes of fast and frugal resistance comes from the troubled U.S. airline industry. Throughout the 1960s and 1970s, the practice of "bumping" passengers from deliberately overbooked flights had produced outrage, lawsuits, and calls for outlawing the practice. Bumping made everyone—the airlines, their passengers, and regulators—increasingly unhappy.

Julian Simon, then a University of Illinois economist, came up with a simple, cheap, and fast experiment to help eliminate the airlines' dysfunctional economics of involuntary overbooking:

> The next day, when shaving, it occurred to me that there must be
> a better way; indeed, a market could solve the problem by finding

those people who least mind waiting for the next flight. The airline flight personnel would simply need to ask each ticket holder the lowest amount he or she would be willing to accept to wait for the next plane, and then select the necessary number of low bidders. The practical details fell into place before the shave was complete. The scheme is simply a reverse auction, asking each ticket holder to write down the lowest amount she or he would be happy to accept in return for waiting for the next flight.

Throughout 1967 and 1968, Simon reached out to virtually every airline proposing a reverse auction experiment. The response that he received from the director of reservations of the (now defunct) Pan American airline, Simon wrote, "was more colorful than most but otherwise not atypical."

> Dear Mr. Simon: Mr. Shannon, our senior vice president-operations, asked me to thank you for your most practical solution to the reservations booking problem. Of course, we instituted the procedure immediately, after having the instructions for bidding translated into 18 languages. We are unable to tell you of the expected excellent results as yet because the first flight on which the system was tried was on St. Patrick's Day, the 17th. The plane hasn't left as of this writing [March 22, 1967], since the passengers keep changing their bids. We feel it's highly unfortunate also that the first occasion for using the procedure should happen on a flight destined for Ireland that day, and some of the passengers were in rather unusual condition! We will keep you advised.

Simon recalls that his proposal was treated with derision by both the airline industry and fellow economists. Indeed, Simon's greatest professional surprise and disappointment came from two respected economists who would both become Nobel laureates:

> But the reactions of the man I consider the greatest economist [then] alive (and the greatest spirit), and another of those economists whose work I honor most, were unusual and therefore particularly interesting. The latter, George Stigler [who received a 1982 Nobel Prize in Economics], wrote that the scheme would not work because the

passengers would form cartels and hold up the airlines for very high prices. "Since your scheme strikes me as intellectually admirable and administratively impossible ... You should explore the possibilities of collusion by a group of forty unemployed people," Stigler wrote.

Milton Friedman [recipient of the 1976 Nobel Prize in Economics] wrote as follows: "If the plan is as good as you and I think it is, I am utterly baffled by the unwillingness of one or more of the airlines to experiment with it. I conclude that we must be overlooking something. I realize that you have tested this quite exhaustively, and I have no reason to question your results; yet I find it even harder to believe that opportunities for large increments of profit are being rejected for wholly irrational reasons."

So not one but two future Nobel economists thought Simon's simple proposal impractical and unrealistic. The irony, of course, is that Simon wasn't calling for a regulatory reverse auction revolution, but merely for fast and frugal experimental tests.

"In all my discussions on the idea," he wrote, "I insisted that one should not decide about it in the abstract, or even on the basis of hypothetical experimental data, but instead should conduct an actual experiment. But I was unable to persuade any airline ... to conduct an experiment for even one day on a single airline at a single airport at a single boarding gate—an experiment that I believed would be sufficient, even with the inevitable breakdowns in any new activity. Rather, the industry and the bureaucrats preferred to insist on the basis of their 'logic' alone that the scheme could not work." Indeed.

Ultimately, regulators effectively required the airlines to explore Simon's proposal in 1978. After rapid industrywide experimentation, these "volunteer" programs quickly became industry norms. Simon's scheme had positively improved both airline economics and the customer experience. Within a few years, the rate of involuntary bumpings had dropped from 6.4 in 100,000 in 1978 to 1.1 per 100,000 by the mid-1980s.

The essential insight for professionals in various industries and Nobel economists alike was stark and humbling: When in doubt, do the simple, fast, and frugal experiment; you'll learn something valuable. If there are no doubts, what truly justifies that impregnable confidence?

Simon's decade-plus saga recalls the wonderful joke about the two University of Chicago economists (both Friedman and Stigler were University of Chicago economists) walking down the street. They spot a $20 bill lying on the sidewalk. One bends down to pick it up; the other economist stops him: "If it was a real $20 bill," he says, "someone would have picked it up already."

Of course, the actual reality—as Simon's story persuasively suggests—is that there are plenty of $20 bills (and $50, $100 and even $1,000 bills) lying around. The business world is littered with them. But to pick them up, you have to be willing to try a cheap experiment or two.

FROM BLOCKBUSTER TO NETFLIX

This chapter began with Blockbuster; it ends with Netflix. Founded in 1997, Netflix was Blockbuster's most successful rival. The company's "No late fees" positioning ruthlessly exploited Blockbuster's greatest weakness. Netflix won both market share and raves for its customer service. Its users loved it. Netflix was also everything Blockbuster wasn't as an innovator. The company cultivated a kaizen culture of continuous experimentation. As Netflix chief product officer Neil Hunt says:

> We test a lot of algorithmic and data-level variations in movie discovery. We explore large and small variations on the recommendations system, including positioning and tools for input of taste preferences, ways to present recommendations, whether or not explaining recommendations drives credibility, etc.
>
> We explore such basics as play and add-to-queue button placement, size, and functionality. We tested the switch last year of whether the home page should be DVD or streaming focused.
>
> ... We are very proud of our empirical focus because it makes us humble—we realize that most of the time, we don't know up-front

what customers want. The feedback from testing quickly sets us straight, and helps make sure that our efforts are really focused at optimizing the things that make a difference in the customer experience.

If I had to summarize our learnings in three words: "Simple trumps complete."

But when Netflix committed to the biggest strategic shift in its history, the company chose to plunge ahead in a manner seemingly in defiance of its self-declared "test everything" culture. In 2011, Netflix began splitting its online streaming video offerings with its DVD-by-mail delivery service. The proposed changes proved confusing and expensive. Customers were outraged. Netflix's stock price dropped by over 50 percent. Reed Hastings, the company's founder and CEO, publicly apologized for how poorly his company had managed the transition.

As John List, one of the true pioneers in experimental economics, and Uri Gneezy recount in their book *The Y Axis,* the cost of Netflix's non-experimentation was enormous. "The company could have avoided the loss of billions of dollars and damage to its brand if it had run some simple field experiments," they wrote. "Rather than coming up with a national scheme to thrust upon customers ... all Netflix had to do was run a pilot of their grand plan in a small portion of the country—say, San Diego—and then study its customers' reactions. The small-scale experiment could have saved the company lots of money without cutting its value. ...

"Even if this experiment had stirred up some negative attention, Netflix executives could have explained it was a local snag. The damage would have been much smaller and the experiment worth its weight in gold."

Do you detect a pattern here? According to List and Gneezy, "When we discuss experimentation with business leaders, they usually reply by saying, 'tests are expensive to run.' After we point out that they are not, we turn the tables on them by showing how expensive it is *not* to experiment. ..."

Exactly. The lesson of both yesteryear's Blockbuster and today's Netflix is that, on average, not running fast and frugal experiments proves much more expensive than running them. Boards should challenge C-suite executives—and C-suite executives should challenge their direct employees—when fast and frugal experiments *aren't* being regularly run. Yes, designing portfolios of simple, fast, and frugal experiments is a creative and operational challenge. But the evidence suggests overwhelmingly that the real-world business benefits of doing so are immensely greater than its risks or costs. Don't resist that message; experiment with it.

8 EXPLORING AND EXPLOITING EXPERIMENTATION: THE 5×5×5 APPROACH

Increasingly competitive global markets have made innovation a top-management imperative. When cost-cutting efforts hit diminishing returns, investments designed to add new value enjoy more serious consideration. Consultants, advisors, and assorted management gurus are called upon for innovation insights. New tools and technologies are considered to renew or redefine innovation processes. Firms revisit the fundamentals of their innovation culture and practices.

While executives declare a greater willingness to innovate, their concerns over costs and risks remain. They're skeptical of innovation transformations that might undermine key customer and partner relationships. They're cynical about innovation incrementalism that promises new value creation on the margins. More firms want to be more innovative about being more innovative. They want to tap their people's collective expertise quickly and cheaply, while preserving top management's strategic innovation prerogatives. In effect, they seek to strike a better balance between bottom-up innovation efforts and top-down strategic imperatives.

5×5×5: RAPID INNOVATION METHODOLOGY

This desire has made many firms more receptive to exploring novel innovation methodologies. The 5×5×5 X-team approach is a rapid innovation methodology emphasizing lightweight, high-impact business experimentation. The term *lightweight* means surprisingly inexpensive in terms of organizational time, money, and resources; *high impact* means that the proposed experiments test business hypotheses that the firm's management deeply cares about. Combining lightweight and high impact inherently commands organizational curiosity, attention, and respect. The lean and agile nature of the experiments ensures that they generate actionable insights remarkably fast. They're less proofs of concept than invitations to take the next innovation steps quickly.

The methodology has been used effectively by global enterprises ranging from European conglomerates to Brazilian media giants to Australian financial services firms to U.S. consumer product companies. What began as a quasi-academic gimmick to goad students and companies away from their infatuation with good ideas became a distinctive innovation option. The appeal comes not only from the 5×5×5's emphasis on speed and low cost, but its explicit effort to align the firm's improvisational talent with top management's articulated vision. The 5×5×5 is designed to employ a firm's cultural and organizational diversity to inspire experimental ingenuity.

The central purpose of the 5×5×5 is to create a vibrant internal market of business hypotheses and portfolios of experiments for the enterprise. Rivalry and competition complement collaborations and cooperation. The goal is to link lean and agile experimentation explicitly to lean and agile innovation. Simple experiments lead to strategic initiatives and impact. Experiment portfolios offer top management the chance not only to explore disruptive opportunities, but to better manage innovation risk. While the experiments and the business hypotheses being tested are valuable, the methodology is also a source of human capital formation. To paraphrase the pioneering French industrial sociologist Le Play, "The most important product of the experiment isn't the data, it's the experimenter." Aligning experimentation with innovation is important, but expanding the boundaries of human capability and creativity around innovation and experimentation is even more so.

5×5×5 DESIGN AND CONSTRAINTS

The 5×5×5 design is simple and straightforward. A minimum of 5 teams of 5 people each are given no more than 5 days to come up with a portfolio of 5 "business experiments" that should take no longer than 5 weeks to run and cost no more than 5,000 euros to conduct. Each experiment should have a business case attached that explains how running that experiment gives tremendous insight into a possible savings of 5 million euros or a 5-million-euro growth opportunity for the firm.

There's nothing magical about the number 5. The point is to insist upon experimental ingenuity within defined constraints. The purpose of these constraints is obvious: Individuals and organizations should come up with experiments offering high-impact potential at great speed and low cost. The constraints are less important than the ingenuity they inspire. The issue isn't whether the euro (or dollar) budget for a proposed experiment exceeds 5,000 or if the experiment could be done over two weekends instead of thirty-five days—it's pushing a small group of people to think "inside the box."

But the 5×5×5 provides an unusual and very special box—one that's provocatively shaped and made from innovative materials. This box is designed to make disruptive innovation not just possible but probable. Should the 5×5 teams do suitably clever jobs of thinking inside this box, they'll come up with portfolios of experiments with the potential and power to transform their organization's innovation culture—if that is what the organization wants.

In larger firms, some twenty-five to thirty-five high-potential managers and workers from across the organization are identified that top management wants to participate in the X-teams exercise. This ordinarily results in five to seven X-teams. They're briefed—with examples, a framework, and suggested process options—on what their deliverables should look like. As a rule, participants get no special compensation—although travel expenses, appropriate time allowances, and other benefits help ensure adequate collaboration with colleagues.

The big motivational and organizational lure is that each X-team will present its portfolio to top management, who ideally include the CEO, a C-suite colleague or three, and occasionally a nonexecutive director and prestigious outsider. These X-teams are fully aware that they are competing with their colleagues to come up with the best possible portfolios to present to their bosses. Rivalry has proven an effective mechanism to focus team attention, energy, and ingenuity on experimental designs that are most likely to impress their superiors. Occasionally, presentations are made before top profit and loss (P&L) executives of the various business groups—the de facto CEOs of the key business units. The 5×5×5 doesn't work unless the participants are confident that their work is being taken seriously by people with the power to either fund the experiments or advance the presenters' careers.

5×5×5 PORTFOLIOS:
WINDOWS INTO ENTERPRISE CULTURE

Simple statistics bolster the odds of creating truly impressive experiments. Five teams times five portfolios equals twenty-five experiments—and six teams times five portfolios equals thirty experiments—and those numbers help guarantee desirable outcomes. While half the proposed experiments—with redundancy and overlap—will be below, say, "median quality," the human capital structure of the 5×5×5 marketplace virtually assures that at least 10 percent to 20 percent of the proposed experiments and hypotheses will be first-rate. The proportion, in fact, is usually higher

There are always—without exception—at least three or four experiments that make top management sit up straight, their eyes widening (or narrowing, depending on temperament), and incredulously ask, "We can do that!?"

Mais, oui—for roughly $5,000, five weeks, and the will to proceed. Every firm should hope—and reasonably expect—that twenty-five or more of its most talented people in collaboratively creative competition with each other should be able to come up with at least three or four truly startling concepts. If they can't, that's important information for top management to know. The absence of innovative experimental proposals is as important for the C-suite to understand as their presence.

More often than not, however, portfolio presentations produce a top tier of world-class experiments. Just below that is a layer of three or four business hypotheses that either individual team members love and want to champion or hold unexpected appeal for a particular business unit that can't wait to try it out (or, in some lucky instances, both at the same time). There are, additionally, the occasional "ugly duckling" experiment that captures the fancy of one or two well-regarded enterprise intrapreneurs. They see—or believe they see—something that elevates the merely decent and intriguing to the level of "this could be the start of something big."

The beauty of well-designed 5×5×5s is that organizations can't help but learn from them. Portfolios become lenses and windows—and mirrors—into an enterprise's culture and priorities.

If twenty-two of twenty-five experiments—roughly 90 percent—focus on customers or clients, then why aren't these participants proposing innovative experiments with their key suppliers and partners? If a portfolio focus emphasizes diversification opportunities, are the 5×5×5 people paying enough attention to organic growth with existing customers? The experiments that aren't being proposed may be as useful and revealing to top management—and the 5×5×5 participants—as the ones that are.

This is a major reason why portfolio concepts are so valuable. It's not enough to come up with 5 individually impressive experiments. How X-teams think through and articulate their focus and philosophy of experimentation is equally significant. Why *these* experiments? What does *this* portfolio say about which innovation opportunities should be identified and explored inside the firm?

Getting talented and ambitious people to take disciplined approaches to these issues can be extraordinarily valuable. X-teams that are forced to think rigorously through portfolio rationales of experiments rather than business plans or proposals give firms innovation pipelines that blend actionable appreciation of risk and reward. By making rapidly business hypotheses the *schwerpunkt* (focus) of innovation design, firms can create a better balance between action and analysis. This frequently represents an important cultural value for innovators. After all, no one can get an MBA from an elite school without performing reams of analysis. But remarkably few world-class business schools perform rapid experimentation and test a curricular requirement.

That's the core concept, core methodology, and consistent core outcome. But that leaves out the elements that give richness and flavor to the individual experimental experiences. For example, many firms put no constraints on the portfolios that they're asking their people to deliver: Anything goes. Blow us away. Other organizations impose a bias or meta-constraint. They want an experiment portfolio to address particular concerns: those of customers, suppliers, Web 2.0, or emerging markets. In other words, top management wants experimentation that aligns with innovation along a particular dimension. Sometimes this additional filter frustrates participants (an interesting finding in itself); other times, it evokes a much higher degree of experimental specificity. Observing how teams conform to—and subvert—what top management says that it wants is always revealing.

RISING TO THE INNOVATION CHALLENGE

The professional development value of the 5×5×5 approach may have been a pleasant surprise, but that shouldn't overshadow the other vital aspect. Virtually no participants—and certainly no members of top management—consider the exercise to be a poor or mediocre use of their time. On the contrary, the overwhelming majority of participants comment that it yields disproportionate value for the time invested. Part of that is attributable to the professional development aspect. But the simple reality is that top management has a relatively inexpensive way to see how innovative and creative its people can be when challenged to come up with business experiment portfolios. Instead of being an intellectual or academic exercise, 5×5×5s turn out to be reality checks for organizations that say they want to be more innovative.

Successful 5×5×5s make people more effective innovators. More effective innovators mean more effective innovations. Faster. Better. Cheaper. Even moderately successful 5×5×5 experiences attract broader attention. The 5×5×5 is a viral innovation methodology. It's infectiously innovative.

The willingness to ask simple questions is essential. The 5×5×5 offers a fast, cheap, and ingenious method for innovators to safely revisit—and test—business fundamentals. Simple questions about customer segmentation, sales, pricing, design, performance, and language successfully inspire high-impact hypotheses. Simplicity invites ingenuity.

The 5×5×5 methodology has worked well in organizations that love to play with ideas themselves, as well as firms that, frankly, would rather draw innovation inspiration from outside consultants. Collaborative firms have welcomed 5×5×5s as a culturally compatible innovation approach while top-down, quasi-autocratic leaderships have cautiously embraced 5×5×5s as a safe, cost-effective diversification of their innovation budget.

9 THE 5×5 PORTFOLIO: THREE REAL WORLD EXAMPLES

The three 5×5 portfolios described in this chapter are composites of experiments proposed by X-teams from companies in three disparate industries. But they fairly and accurately represent the breadth and depth of hypotheses and experiments effective X-teams explore. The individual firms have graciously allowed me to draw upon their X-teams' efforts to share what successful portfolios look like.

These portfolios are best viewed less as idealized models and more as templates for context and reference. Several selected experiments no doubt will feel as anachronistic as Sir Isaac Newton's apple. Others are more contemporary in feel. Collectively, they convey the sweep, scope, and ingenuity that motivated X-teams bring to their challenges.

When they were presented to their top managements, the business hypotheses and experiments were genuinely seen as provocative and subversive. Top managements—and the X-teams themselves—were palpably surprised by what they had created.

Instead of feeling constrained by the time, budget, and resource limitations, the X-teams found a focused freedom. They never had done this sort of design, analysis, or presentation before. The novelty was appealing.

According to my notes and available recollections, not a single experiment or hypotheses presented here popped out of an X-team's first meeting. They emerged from iterative argument and interaction. They were products of collaboration.

This is a biased sample. Honesty requires disclosing that these are "above-the-median" portfolios. That is, they reflect successful outcomes. Poorly designed and presented experiments are discussed elsewhere.

Including the mediocrities seemed pointless. From my observation, truly mediocre experiments and hypotheses are identified and dismissed quickly. (Unless the X-team itself is mediocre ... which is a different problem.) Serious portfolio discussions typically create a "flight to quality." That is, very few 5×5 teams care to spend their time and efforts on hypotheses and experiments that don't much matter.

But honesty also compels disclosing that none of the best experiments are in these mock portfolios. For competitive and proprietary reasons, no company agreed to share their best experiments publicly. But they did allow me to describe edited versions of their bronze and silver medalists.

Consequently, neither borderline nor best-in-class ideas are fairly represented here. But these portfolios reveal what organizations thought was quite good at the time. Just as important, they offer practical frames of reference for innovation leaders and managers to assess their own hypotheses and experiments. Innovators surely wouldn't want to do worse than the experiments discussed here— but they should be confident that it's demonstrably possible to do much better.

PORTFOLIO 1: THE CREDIT CARD COMPANY

One feisty credit card competitor took pride in being technologically adept. But the firm recognized that it needed to improve customer service. Fraud and security issues, however, were a constant concern.

This X-team's portfolio focused on improving the cardholder's experiences with customer service while adding new value to existing card features. To diversify the X-team's portfolio, one hypothesis and experiment were dedicated to improving security.

Experiment 1

The Business Case: Customers phoning the company for assistance wanted faster, better, and more helpful interaction with call center staffs. Instead of simply adding more documentation for call center representatives to search through, why not experiment with instant messaging (IM) and online chatting to facilitate information sharing? Observing IM and chatting could provide useful data for both training and more streamlined customer service support.

The Business Hypothesis: If we put IM/chat capabilities on our customer call center systems, our representatives will share queries and information faster, thus responding to customers faster and with more accurate information.

The Experiment: Give IM/chat software to about forty people in two different call centers. Do basic, but not extensive, training. Offer small bonuses for the most creative and effective uses of chatting to serve customers. Compare customer response times, customer satisfaction levels, and representative satisfaction levels with non-IM/chat performance benchmarks after four weeks. Observe if particularly effective patterns of IM/chat usage emerge.

Experiment 2

The Business Case: Unhappy customer callers want some sort of resolution or compensation for perceived errors and mistakes. Customer service representatives can remove disputed charges or waive fees temporarily. But many customers are unhappy that all this does is return them to the status quo. Is there some way to give legitimately unhappy customers something of value that will mollify them while building greater measurable customer loyalty?

The Business Hypothesis: If we give legitimately unhappy customers small but reasonable numbers of reward points from our loyalty program, their customer satisfaction and loyalty will measurably increase, while their unhappiness will decline. The ability to give points also empowers customer service representatives. Can we use our points program to innovatively address customer dissatisfaction?

The Experiments: Do a quick internal web survey to learn more about the nature of complaints and customer dissatisfaction. Based on those preliminary findings, work with a group of fifty or sixty customer service representatives to segment customers and complaints better.

- Give twenty customer service reps a budget of points that they can give, at their discretion, to qualified complainers during their shifts.

- Require twenty customer service representatives to compensate a randomized mix of complainers based on a per-complaint formula.

- Have twenty customer service representatives ask qualified complainers if being given a set amount of points would help mollify them. If yes, give them the points.

- Have a randomized mix of complainers via the Internet be sent a set amount of compensatory points.

- Do a mix of email, text message, and outbound customer center calls 48 hours, one week, two weeks later, or all three to assess the customer satisfaction level. Monitor the loyalty program accounts to see if there is any statistically significant change in customer behavior within 90 days. Survey the customer service representatives to see which approach gives them the greatest sense of empowerment and effective resolution.

Experiment 3

The Business Case: Customer swipe machines for debit and credit cards are becoming ubiquitous at supermarkets, stores, and fast food outlets. The screen and swipe combination effectively processes every transaction, but it also offers a "touch point" where the card company can add value for the customer. Could these machines be used to enhance the card user's experience?

The Business Hypothesis: If we put relevant messages, offers, or both on the screen during a transaction, we can both constructively influence both merchant and customer behavior and make our card(s) the preferred choice for transactions.

The Experiments: Explore if we have point-of-sale (POS) relationships or merchant relationships who would partner for small-scale experiments.

Determine a suite of "swipe screen" message prompts to test. For example, when a purchase exceeds a certain level—say, $25—a "Would You Like a 5% discount?" prompt would be displayed. Other prompts could be used to suggest purchases or promotions, offer reward program points, ask if the customer wants an immediate email receipt, and other incentives.

Can we collaborate with the store(s) to come up with mutually beneficial messages and prompts? Are there prototype swipe POS devices whose future design and deployment we can influence?

Can we tailor messages and prompts as a function of debit or credit card selection?

Experiment 4

The Business Issue: Credit card statements and receipts are increasingly being sent via email. Are there ways to turn these digital documents into dynamic coupons, promotions, and advertisements for both our customers and merchants?

The Business Hypothesis: If we integrate advertising and promotional offerings into our billing statements and online receipts while adding links to merchants, we can drive traffic to favored merchants and nonintrusively offer our cardholders special deals through their statements. Merchant partners might pay us for promotional positioning on our statements; customers may be happy—or at least less unhappy—to get and click open their credit card statements.

If only 2.5 percent of our 40 million cardholders click on an embedded advertising or promotional link per billing cycle, that's 100,000 clickthroughs a month and over 1 million a year.

The Business Experiment: If we embed a variety of links into our online billing statements, what percent of customers will click on them? Should we segment who we test, or should it be a cross section of online bill payers? Send out a randomized set of 50,000 billing statements with embedded links to merchant partners.

We can test one, two, three, four, and five embedded links and compare the results. We can test coupon offers against discount offers against "exploding coupon" offers (i.e., click within the next ten minutes or lose this one-time discount …).

Should we test charging our merchants for the privilege of these digital promotions, or should we test pay-for-performance? Should we handle email receipts differently than we do statements? Should we test sponsored receipts—i.e., "This email receipt is brought to you by Intuit personal finance software"?

Experiment 5

The Business Issue: Credit card fraud and identity theft remain serious problems for all of us. Anything that we can do to validate identity rapidly at the point of transaction can improve security for the cardholder, merchant, and issuer. The rise of mobile devices—from cell phone to iPhone to BlackBerry—has transformed how cardholders communicate. Can these devices be easily, cost-effectively, and reliably integrated into our security and validation infrastructure? Deterring or reducing even 3 percent of known fraud would save us literally millions of dollars.

The Business Hypothesis: If we invite people to register their mobile device accessibility with us, we can explore real-time ways to reach them if there are "point-of-transaction" security or identification issues. By better utilizing our customers' devices, we can better protect everyone in our value chain.

The Business Experiment: Write a customer service representative script that asks a randomized selection of callers if they carry and use mobile devices with texting, email, or both. If yes, ask if they would be willing to share their mobile address so they can be contacted instantly if security questions arise. Track acceptances and declines.

For a randomized subset of callers, offer a points or annual fee reduction inducement to share their mobile addresses.

For customers who pay online, offer a randomized selection of bill payers a button to click that would allow them to register their mobile addresses.

Bonus experiment: Should we ask customers if their devices have global positioning system (GPS) locators? Should we ask them for waivers so we can see their actual locations when they use their cards? This would be a nonintrusive way of matching cardholders with their card usage.

PORTFOLIO 2: THE PHARMA COMPANY

A relatively young and entrepreneurial pharmaceutical company—which had partnership agreements with larger pharma firms—wanted its X-teams to invent experiments that kept the users in mind. This composite portfolio focused on the importance of compliance—also known as *adherence*—for prescription users. Innumerable surveys confirm that almost half of older patients don't take their medications in the right quantity, frequency, or schedule that their doctors prescribe.

People take their medicine twice instead of three times a day, or they double the dosage if they forgot to take an earlier pill. Compliance concerns are particularly serious among the elderly, who often have to track several prescription regimes. Compliance/adherence is a multibillion-dollar business and medical issue.

Experiment 1

The Business Issue: Pharma companies typically play little to no role in monitoring or facilitating compliance. Doctors and pharmacies prescribe and distribute the drugs to patients. Gaining greater visibility into patient and caregiver concerns around compliance and adherence could both help promote the medical efficacy of the drugs and potentially blunt the legal exposures that the firm might encounter. Pharma companies are understandably concerned that U.S. and European regulators could impose compliance-related rules that could dramatically increase the companies' costs and responsibilities.

Enhancing compliance, helping reduce compliance-related side effects, and minimizing legal risks and liabilities potentially could be worth tens of millions of dollars to the firm.

The Business Hypothesis: If we invite prescription holders to call us to discuss their compliance with/adherence to taking our medications, a significant number will respond. If our users don't respond to a compliance conversation, that's also useful information. Learning how many—and who—respond to these invitations is useful as well.

We could play constructive roles in educating our customers and aiding compliance. This effort would give us greater credibility with the medical community, insurers, and regulators, and at low cost.

The Business Experiment: Pick a drug that we particularly want to know about in terms of compliance. Identify a representative sample of pharmacies that distribute it. Develop two test brochures inviting users to call a toll-free number to discuss compliance-related issues. One brochure might emphasize getting advice on compliance; the other might emphasize the importance of improving compliance. Have the pharmacies distribute one or the other brochure with the prescription over a two- or three-week span.

Track the number of calls and relative response rates of the brochures. Have the call center ask a short set of questions identifying the callers and learning about their compliance concerns. Ask if they would prefer to receive the medication through a patch instead of via a pill or other method.

Experiment 2

The Business Issue: Compliance has many components. For reasons of both price and dosage, pill-splitting has grown in popularity—particularly among older populations. Tablet formulation processes and pricing practices mean that—for many kinds of pills—cutting them in half may make both medical and economic sense. This practice has enormous implications for production, marketing, and pricing. Is pill-splitting something that should be ignored, encouraged, or discouraged as a customer behavior?

The Business Hypothesis: If we encourage our users to split pills (if medically appropriate), we can improve our brand as a responsible, compliance-friendly, and consumer-friendly pharmaceutical manufacturer. In addition, we might gain insight into how involved certain customer segments are in customizing and personalizing dosages. What's the mix between pill-splitting for medical versus economic reasons?

The Business Experiment: Identify which of our pills are split the most and why. Identify representative pharmacies and health maintenance organizations (HMOs) that disseminate them. For the two medications that are the best candidates for splitting, develop poster and handout material asking, "Does Your Doctor Think You Should Be a Splitter?" The material should feature a picture of a high-end pill-splitter with our pills neatly cleaved.

Invite prescription customers to call a toll-free number to get 50 percent, 25 percent, or 75 percent off the list price for the premium pill splitter. Have a statistically significant number of offers made both at the representative pharmacies and HMO offices. Track relative response rates. Procure the names of physicians on calls. Call an appropriate sample of prescribing physicians to learn their attitudes and recommendations toward their patients who split medications.

Optional follow-up: Ask how the pill-splitters were actually used, whether they influenced the number of pills purchased, or both.

Experiment 3

The Business Issue: The focus remains on compliance/adherence. The rise of mobile devices suggests both a medium for real-time notification and drug dispensing for compliance. Can we use mobile communications devices to facilitate and enhance compliance/adherence?

The Business Hypothesis: If we link pill dispensers with mobile communications devices, we could notify people to take their pills in a timely fashion. This helps us improve the medical efficacy of the drug while expanding our knowledge of how people actually adhere to prescription protocols.

The Business Experiment: Identify which of our medications have the most serious compliance issues.

Seek 100 to 200 volunteers who take those medications and claim to have compliance concerns. Take their mobile phone numbers, or give them a pager with a pillbox attached. Take their prescription information and have a leased server page or call their devices to remind the patients to take their medicines.

Compare the compliance rates of the participants with a random sample of nonvolunteers. Get responses from experiment participants—and their doctors—as to effectiveness of the tele-reminders.

Experiment 4

The Business Issue: Elderly people in particular have compliance/adherence challenges. Ensuring and tracking pill consumption is particularly difficult for older patients who require external nursing or supervision. Alternative drug delivery mechanisms—notably transdermal patches—may be a better medium to assure compliance/adherence and medical effectiveness.

Changing the patches every two or three days may be more comfortable and effective than taking pills two or three times a day at specific times. Therefore, switching from pill to patch delivery platforms could improve compliance dramatically in elderly patients—particularly those in assisted living or community facilities.

Given that aging populations are growing significantly in the industrialized world, transdermal drug delivery for the elderly represents a huge market opportunity.

The Business Hypothesis: If we test the logistics and appeal of transdermal delivery in the elderly, we can better decide what investments to make in reformulation and marketing for that growing user segment. We think the best way to test that hypothesis is not to target the aging directly, but to work with and through health care delivery personnel.

The Business Experiment: Find a representative sample of nursing homes, gerontology wards, and assisted living facilities. Procure individual and institutional waivers for placebo testing. With appropriate control groups, test how both individuals and their nursing support staff react to placebo patches administered daily, once every two days, or once every three days. Compare compliance attitudes and behaviors to traditional tablet consumption. Do nursing staff and their charges prefer pill or patch adherence regimes over a two-week span?

Experiment 5

The Business Issue: Because pharma companies have typically focused their efforts on drug development, marketing, and distribution issues, they don't fully appreciate the strategic efficacy issues associated with compliance/adherence. In an evolving legal, regulatory, and technological environment, getting the organization to understand compliance better becomes important.

We think compliance/adherence is becoming a central challenge for the pharma industry and how it is perceived, both as a business and as a regulated entity. We need to take quick, high-profile, but

cost-effective steps to promote a greater internal focus and competence around the future of compliance/adherence and medical
efficacy.

The Business Hypothesis: If we recognize and reward creative
experimentation around compliance, our key people will bring a
greater compliance sensibility to their marketing, development,
regulatory, and channel management decision making.

The Business Experiment: Hold a web-based competition with
prizes to recognize and reward the five best compliance-oriented
business proposals. Have the contest open to everyone in the firm,
but also organize a handful of 5×5 X-teams to generate recommendations collaboratively. Compare the quality of the team-generated
versus open-submission suggestions.

PORTFOLIO 3: DISPOSABLE DIAPER COMPANY

This firm, a maker and seller of disposable diapers, wanted to
rekindle its innovation ethos. International competition—notably
from Japan—was setting the innovation pace for this multibillion-
dollar industry. There was no shortage of good ideas, but approval
processes and testing protocols invariably took more than a year
before internal champions could begin to procure senior management support.

Declining margins and fears that the brand's innovation image
was at risk created an impetus for change. While respecting process
and rigor, the firm's leadership declared a willingness to accelerate
testing. So long as no appreciable costs or risks materialized, rapid
experimentation was a welcome spur to getting more people more
innovative faster.

Given the enterprise brand orientation, the emphasis of one
X-team was not innovating better diapers, but innovating better
diapering experiences.

Experiment 1

The Business Issue: Extensive research has suggested that a significant percentage of parents change disposable diapers at night, in the dark. These parents don't want to risk waking the baby by turning on the light. Innovating around the brand attribute of "keeping your baby dry and asleep" has become a design challenge.

The perception was that if parents or other caretakers could change diapers more easily in the dark, they would choose to do so. Elevating the challenge from having a user-friendly diaper to changing diapers in the dark easily without waking the baby, the firm could differentiate its branding and new value message meaningfully.

The Business Hypothesis: If we go "beyond the diaper" to light the nursery, the crib, or both in a new way, we can make the nighttime diaper changing experience faster, easier, and less intrusive.

The Business Experiment [A]:   An X-team member had come across a cheap "Made in China" night-light with a remote-controlled rheostat (a light dimmer switch). The team suggested finding new parents willing to volunteer to test if remote-controlled dimming, with the ability to set a level bright enough to change a diaper without disturbing the baby, would interest them.

The X-team suggested in-store experiments. The night-light could be bundled in with the diaper package, or the company could issue a coupon for in-store redemption. Alternatively, stores could hold an expert seminar for promotion to demonstrate the concept.

Would the offer sway rival disposable-diaper purchasers? Would it shift shoppers' perceptions of brand leadership and innovation assistance?

The Business Experiment [B]:   Use very-low-luminosity, "glow-in-the-dark" adhesive tape to create diaper tabs to made night changes easier. Test the redesigned diapers with both men and women. Explore different configurations, locations, and luminosities of the tape. Could this be a patentable mix of proprietary material, design, and the production process?

Experiment 3

The Business Issue: Parents of newborns are obsessed with their child's health. They're concerned about sniffles, sneezes, burps, and other various emanations. They worry about calling the doctor or waiting until morning. They're prepared to do almost anything they can to gain greater insight into their child's health and well-being.

Parents constantly changing diapers also are constantly concerned about whether the baby's eliminations reveal any warning signs about his or her health. The consistency and color of excretions and fluids may signal something important ... or not.

Perhaps changing a diaper shouldn't just be a cleaning- and hygiene-oriented experience, but a diagnostic and health-affirming opportunity. Could disposables become diagnostic tool for parents? Are there materials that could be put in a diaper or used in conjunction with a diaper that might aid parents in diagnosing any significant health issues with their babies? Turning disposable diapers into diagnostic tools would be a multibillion-dollar game-changer for the business.

The Business Hypothesis: If we offer low-cost, decent-quality diagnostics with our disposable diapers, a significant proportion of parents would use them eagerly and gratefully to monitor their babies' health. In the same way that home pregnancy tests became a market-pioneering category of home diagnostic, disposable diaper diagnostics can create a compelling new value proposition for parents worldwide.

The X-team believes that there is enough neonatal care expertise and clinical laboratory knowledge that certain gastrointestinal ailments, kidney issues, or both could be identified with inexpensive, mass-produced materials. Should there be dedicated tests for specific issues; or should we test for multiple conditions?

The Business Experiment: Seek volunteer parents who are willing to collaborate with us on trying various test configurations for diagnostic diapers. Should there be "diagnostrips" that are built into the diapers? Alternatively, should diagnostic wands be bundled into the diaper pack to offer effective diagnostic systems? What diagnostic displays are preferred: Color? A plus or minus sign? Something else? What minimum level of confidence or concern do parents need to reach to keep them from calling a doctor?

As with the remote control night-light, how might in-store demonstrations of the value proposition influence experimental designs and prototypes for diaper diagnostics?

Experiment 4

The Business Issue: The diapering experience is messy by its very nature. Babies that have soiled themselves need to be cleaned. Many parents and other caregivers carry around boxes of baby wipes and, perhaps, a Diaper Genie to clean up after their babies. But these items can be both cumbersome and burdensome. Caretakers taking newborns to the park or public areas want convenience. They want lightweight ways to handle diaper changing, cleaning, and disposal.

The Business Hypothesis: If we bundled baby wipes and disposal baggies into travel-light diaper three- or four-packs, people would pay a premium for the ease, packaging, and convenience. This could be an important subcategory of diaper design and packaging. In addition, these travel packs could be bundled into regular, larger-volume disposable packages.

The Business Experiment: Develop and test prototype samples of disposables with both "wash 'n' dry" wipes and throwaway "twist 'n' close" baggies attached.

Procure volunteers to travel with the lightest possible "convenience configurations" with their babies. Ask them to record— with camcorder and audio—their experiences using the materials.

Try to do another off-the-shelf bundling of wipes and baggies. Are there baggies with talc, disinfectants, or both? Should the wipes be placed and attached inside the bag? Put in a special diaper pocket? Should there be per-diaper packaging? Is the better ratio 3:1 or 4:1?

Experiment 5

The Business Issue: Mothers and other caretakers often multitask as they look after babies and toddlers. They are often on the move, attempting to run errands, remember calls to make, and other activities. By nature, the focus of their attention is on the child and when the child might need to be changed. To keep track of everything, the caregivers often make extensive lists, which are modified and edited throughout the day. Perhaps the diaper—which has to be attended to regularly anyway—is a good platform for making notes.

The Business Hypothesis: Disposable diapers make an ideal platform for sticky notes, messages, and reminders. If we develop a pressure-sensitive, brightly colored paper, caretakers might be happy to write notes about things that they otherwise might misplace or forget to do. Attach a couple of these pressure papers per diaper as a "Mother's Helper."

The papers wouldn't have to be attached to the diaper; like regular sticky notes, they could be attached in any handy location. The disposable diaper pack would be associated with organization, not just baby care. The regular task of changing the baby would ensure that opportunities for reminders would occur fairly often.

The Business Experiment: Attach different colors, shapes, and sizes of existing sticky notes—and pressure-sensitive sticky papers—to diaper samples. Find volunteer mothers and other caregivers who would be able to use their mobile phones to photograph and verbally record how—or if—they use the notes. (Important: Mothers and other caregivers ordinarily don't like to have mobile devices too close to their babies and infants for hygienic and functional reasons. A hypothesis worth testing is that sticky pressure paper may be a preferable medium to anything electronic when dealing with these young children.)

Over the span of a week, do "diaperized notes" become a part of the changing experience? Are they seen as a useful convenience? What does the content of the notes tend to be? Do different shapes, sizes, and colors of the paper influence its use? Where on the diaper are the papers best attached?

10 KEY STEPS TO THE 5×5

A simple and straightforward ten-step program describes how to bring the benefits of 5×5 X-teams to your organization. The methodology's been tested in entrepreneurial and established organizations all over the world. Flexible, versatile, and powerful, these principles provide high-probability paths to measurable success. The results—experimentally and organizationally—will impress.

The spirit of the guidelines is more important than the letter. Five key phrases best capture that spirit:

- **Make compelling business hypotheses simple and simple business hypotheses compelling.**

The business potential of your hypotheses should be easy to understand. They address fundamental business questions and concerns. The importance of these hypotheses needs to be obvious, and they should invite and excite curiosity. Ideally, they will spark innovation conversations that your organization needs to have. People want to see what experiments you'll devise to test them.

- **Seek to perform simple experiments fast.**

Make sure experiments are accessible and explainable; no MBA or PhD necessary. Impatience is a virtue Most 5×5 experiments take a few weeks to run, but a few may take only a few days, or even a few hours. Could they be more complex, comprehensive, or sophisticated? Of course—but avoid that trap. The 5×5 mission is designing experiments that generate remarkably useful information remarkably fast. Speed and simplicity are reinforcing virtues.

- **Seek to perform fast experiments cheap.**

Speed and simplicity are important. So is cost. Your experiments are lean, mean, and cheap. Very cheap. In fact, they couldn't be done much cheaper. Their cheapness should inspire admiring disbelief and compliments from everyone who sees them. The single best reason for spending more is to make the experiments simpler, faster, or both.

- **Seek results that speak for themselves.**

Simple and fast experiments should yield clear and compelling results. Minimize confusion. Avoid nuance, subtlety, ambiguity, or shades of gray. The outcomes make obvious what should—or shouldn't—be learned next. Whatever the results, they remind and reaffirm why the business hypotheses are important.

Here's an example: An Internet news site made viewers sit through a 12-second advertisement before allowing access to the desired content. Surveys and reviews showed that this alienated many site visitors. The service devised a simple, quick, and clear advertising experiment. The site displayed two advertisements and asked visitors to choose which one to watch before proceeding. This experiment not only provided useful information to advertisers, but it evolved into an advertising format.

- **The best, most influential, and most important products of simple, fast, cheap, and compelling experiments aren't the results, but rather the experimenters.**

Developing insightful business hypotheses and innovative experiments is key. Developing insightful and innovative people, however, is even more important. The 5×5 methodology is more about investment in human capital and capabilities than it is about innovation and experimentation.

Successful 5×5s make people more effective innovators. And more effective innovators mean more effective innovations. Faster. Better. Cheaper. Even moderately successful 5×5 experiences attract wider attention. The 5×5 is a viral innovation methodology. It's infectiously innovative.

Credible outreach is essential to attracting attention and support. Don't overpromise. Rein in grand ambitions. Favor a sharper focus. Emphasize the immediacy and diversity of innovation opportunity. Stress the appeal to talent and ingenuity. You're not selling high-end, high-priced, high-cost strategies. You're enabling high-impact, low-cost innovation. Honoring the 5×5 guidelines guarantees high-impact deliverables.

Begin where people are, not where you would like them to be. Situational awareness is essential. Evoke enterprise issues and themes that "everyone knows" are important. But avoid taking the obvious for granted. The obvious—organizational assumptions and routines that can be challenged usefully—may be remarkable resources and reservoirs of experimental opportunity.

For example, an outdoor sporting gear retailer focused its sales training on identifying individual shoppers predisposed to learning about the store's products. Everyone knew that the training was successful. But one 5×5 team observed that couples and families dominated weekend traffic. Their emergent hypothesis challenged the store's sales mindset: should salespeople learn techniques for selling to couples as well as individuals? This led to suggested training experiments and price promotions targeting couples, including two-for-one offers.

The 5×5 shouldn't seem threatening: make it a business experiment about business experiments. Perhaps paradoxically, the methodology works best when participants aren't trying to be renegades, subversives, or revolutionaries. Disruptive innovation is a byproduct of the methodology, not a goal. The 5×5 gives voice and visibility to disruptive innovation without explicitly celebrating it.

A willingness to ask simple questions is essential. The 5×5 offers a fast, cheap, and ingenious method for innovators to safely revisit—and test—business fundamentals. Simple questions about customer segmentation, sales, pricing, design, performance, and language successfully inspire high-impact hypotheses. Simplicity invites ingenuity.

Persuasive power comes from the disciplined conversion of simple and important questions into insight-generating experiments. Action trumps analysis. Producing actionable insights far faster and cheaper than anyone thought possible is the first measure of effectiveness. You know you're winning mind share when business opportunities are framed not as good ideas or problems to be solved, but testable hypotheses. That's how innovation cultures change.

1. KNOW WHY

The first step is knowing why your organization should explore or embrace the 5×5 X-team approach. Is your "why" the most correct, most persuasive, most attractive, or most important of its kind? Is it the "why" that will attract support from top management? Human resources (HR)? Line managers? Will your "why" avoid or minimize resistance? Whatever your answer is, how do you know it?

If colleagues worry about their organization's sluggish innovation tempo, they're frustrated by resource constraints and time-consuming calls for more detailed analysis, or if talented people feel that their creative energies are underappreciated, then essential ingredients of the "why" are there.

The right "why" also materializes when top management demands employees do much more with much less. It's there when executives insist that business units anticipate customer needs faster. The "why" is there when leadership says that it's serious about innovation—so long as it's not too risky or expensive.

Most organizations have many "whys." Is there an overwhelming reason to do a 5×5? Are there several good ones? Which reasons excite the right people? Which reasons might offend the wrong ones? Pick those that will get colleagues to ask, "Why not?"

If you can't write a 200-word mission statement explaining why your organization needs 5×5 experimentation launched within the next 100 days, then you're not ready to champion it. Authentic "whys" attract authentic support. Like any good invitation, your mission statement should make people want to sign up and bring along a significant other.

2. DECIDE HOW HIGH TO GO

Direct access to the CEO, CFO, or another C-suite executive is terrific. Indirect access is also handy to have. But successful 5×5s don't depend on the top of the pyramid. Higher isn't always better. So long as 5×5 support—or acquiescence—comes from a business unit leader, an enterprise function executive, or a process owner, the methodology can work.

Pick a domain where a 5×5 can have the greatest impact in the least time. Pick a business unit or function where a 5×5 will enjoy the greatest organizational support. Should marketing or customer support have its own 5×5? Perhaps the information technology (IT) group could convene a cross-functional 5×5 with an Internet/intranet emphasis.

HR could sponsor 5×5s under the aegis of professional development and training. Finance might facilitate 5×5s to encourage new thinking around cost-effective innovation. Who would be most supportive? Who would be most receptive? Which leaders have reputations, needs, or desires to champion low-cost, high-impact innovation opportunities? Who has the greatest incentive to innovate faster, better, and cheaper? Are *you* that person?

The CEO isn't essential for 5×5 success, but a senior management commitment to listening and critiquing 5×5 portfolio presentations is indispensable. This methodology can't ignite the enterprise unless real dialogue exists between the X-teams that present it and the leadership that reviews it. Ideally, the CEO is involved directly in the 5×5s. But having the reviewer of the method be the executive who runs the most profitable strategic business unit can work, too.

For example, the CFO-turned-CEO at a mid-sized financial firm had little interest in promoting an internal innovation agenda. He preferred to have external strategy consultants suggest acquisition targets. So he dismissed a 5×5-like proposal from a senior intrapreneur in the HR department.

But two of the company's best-regarded business unit leaders—units that were responsible for over 40 percent of the firm's profits—were very interested. The entire company wouldn't do a 5×5, but three business units would. Their leaders wanted to build internal brands as open-minded innovators. They wanted to invigorate and empower their high-potential people, and the 5×5 offered a swift, resource-effective way to do that.

And 5×5 evangelists don't have to start—or finish—at the top. I've seen successful exercises in small divisions win the CEO's attention, and one-shot, enterprisewide 5×5s have propagated and emulated in multiple business units. The methodology is remarkably versatile and adaptable. Size doesn't matter. Go as high in an organization as you need to go to have the impact that begins to make a difference.

3. PICK A PATRON

Even CEOs need supporters and support who can cut through nonsense and administrivia. If you don't have the CEO on your side, you'll need someone who will—if necessary—send the emails and make the calls that get the right people involved.

You are not asking any of these people for time or money. You want credibility and support. This is someone who deservedly commands respect inside the organization and has earned a reputation for recognizing talent and opportunity. This person has asked for your help because he or she is busy.

The CEO is interested in being your 5×5 patron for reasons that have little to do with organizational politics. It's because this person is genuinely interested in what the results will be. The CEO doesn't necessarily want active involvement, but rather to intervene selectively. This prospective patron wants to be seen for who he or she is: a leader who uses position and influence to make positive things happen. This person gives good advice and recommends good people, saves you from making stupid mistakes, and allows you to tap unexpected resources. (Whatever you do, though, don't ask this person for money.)

This individual may be an HR professional or a retired executive who still comes into the office twice a week. Maybe someone who is an executive's fast-track chief of staff or the business's owner in-law. Perhaps it's an external consultant who has cultivated healthy relationships with the firm's movers and shakers. But the primary motivation here isn't about money. Getting credit is a secondary concern as well. This sponsor will want his or her minimal efforts to have maximum impact, and want to be associated with an unusual—rather than typical—success.

Find this person. Better yet, find two. When your prospective patrons see the 5×5 X-teams' emerging portfolios of experiments, they'll thank you for inviting them to help. You will—correctly and gratefully—acknowledge that it couldn't have been done without them.

4. IT BEARS REPEATING: DON'T ASK FOR MONEY

The 5×5 isn't about financial capital; it's about human capital. Don't ask for money. Ask for access—to people, to time, to technology. Ask for meeting space. Ask for videoconferencing help. Ask for HR to provide a facilitator. Ask for anything that can help make 5×5 X-teams successful. But *don't* ask for money.

Everyone asks for money. Don't be everyone. Be different. Be the innovation initiative that succeeds without a budget. Nothing makes a high-value, low-cost 5×5 proposition more credible or authentic than the fact that it will need minimal resources. Nothing says more about collaborative enterprise and commitment than the ability to deliver great value for small sums.

The absence of money liberates. It frees X-teams to focus on their hypotheses, their experiments, their presentation, and their members. The 5×5 methodology is about creating value. Money comes after that, and recognition and satisfaction come during.

People's time and ingenuity are the most valuable ingredients of this process. Time management is the key organizational enabler for X-team success. Time competently managed ensures that the outcomes will be very good. If time is well managed, the results will be terrific. So I'll say it yet again: Don't ask for money.

5. DEFINE THE SCOPE

What are the measureable and meaningful desired outcomes? Anything? One big thing? Something in between? The 5×5 methodology doesn't care—but you must. Manage expectations by sculpting the scope.

Review your 5×5 mission statement. Some organizations want 5×5s to offer the broadest sweep of innovation insights and opportunity. X-teams are encouraged to create portfolios without boundaries—that is, hypotheses and experiments addressing any issue that they deem important. Nothing—neither price, new media, new materials, new customers, new technologies, nor new sales tactics—is off limits. Anything goes.

Other organizations value constraints. Their X-teams have explicit missions or briefs. Their 5×5 focus is producing portfolios around specific challenges or themes, to wit:

- How should we use new media better?
- How can we collaborate better with key suppliers?
- What experiments should we be doing with our best customers?
- How can we segment our customers and clients better?
- Can we wrap new services around our products?
- How can we make our customer support centers innovation platforms?

Organizations can tailor 5×5s to key functions—marketing, sales, IT, training, and so on—or key processes.

Successful X-team initiatives are organized into one of three categories:

- Open
- Thematic
- Specific

The open 5×5 is aptly described by the "anything goes" philosophy. Top management—and the organization—are open to whatever business hypotheses and experiments that the teams devise. Openness and diversity drive team creativity and opportunity. The X-teams set their own innovation priorities for the enterprise. This gives top management—and the organization—a clear window into what X-team talent believe is most important for the future.

Thematic 5×5s ask X-teams to address specific themes that the organization deems important. Themes can be as generic as "Customers" and "Customer Service" or as technical as "Social Media," "Search," and "Cloud Computing." Top managements asking 5×5 X-teams to hypothesize around themes are generally looking for innovation road maps. Rather than hire outside experts, they want cross-functional teams to provide strategic direction in a poorly understood domain. General Electric's embrace of Sig Sigma quality initiatives in the 1990s offers a model of thematic innovation and experimentation.

Specific 5×5s are controversial, constricting, and not uncommon. Instead of wanting X-teams to challenge top management thinking, specific 5×5 X-teams are challenged to solve explicit problems. In effect, X-teams become innovation SWAT and Special Forces teams with explicit deliverables. One example could be: How can we get more value from our enterprise resource planning (ERP) software?

Prepare for scope creep and mission drift. X-teams often find that a marketing 5×5 mission leads to business hypotheses that tread on sales turf, IT's jurisdiction, or other parts of the business.

Seemingly simple marketing hypotheses and experiments cut across the entire firm. Of course, that also can be part of the benefit.

Conversely, "anything goes" 5×5s occasionally confront top management interventions that push them to address the now. They want the X-team focus narrowed to generate hypotheses and experiments for particular market segments or channel partnerships. Or maybe they want portfolios designed to invigorate a lagging business unit.

Managing 5×5 expectations means being prepared to manage those emerging contingencies. But getting buy-in on the initial 5×5 scope makes managing these expectations easier.

6. PICK A DELIVERY DATE AND DEADLINE

Deadlines and finish lines create intensity and focus. Will your initiative be a true 5×5, where X-teams have only five days to design and develop their experiment portfolios? Will they have two weeks? Four weeks? Five? (Anything longer than five weeks betrays the methodology's ethos of speed and urgency, though.)

The window of opportunity should not be generous. Participants should feel stress. They should feel compelled to make every meeting count. They should use email, Facebook, or whatever other collaborative tools they have at hand to manage their collective effort. They need time to rehearse, to elicit comment and review, to challenge each other. The team members' ability to get the best out of each other's time and effort will be as important as the hypotheses that they generate and the experiments that they design.

7. GET THE EXECUTIVE PRESENTATION COMMITMENT

Putting the executive review meeting in place—and making it work for you—matters most. The meeting makes the 5×5 real. It's the deadline, the finish line, and the entire show. This meeting is where the deliverables get delivered.

X-teams need to engage with senior management around their deliverables—the business hypotheses, portfolios, and experiments. Management needs to make and honor the commitment to meet and review the X-team's proposals. This meeting signals that rapid experimentation merits management's time and attention. The X-teams must be ready to present; top management must be prepared to react and interact.

Presentation meetings typically take a half-day—from 8:30 a.m. to 12 p.m., or 1:30–5 p.m. Five teams typically take between 30 to 40 minutes to make their presentations. A pre-presentation review to identify and manage redundant or overlapping hypotheses and experiments facilitates more time for discussion. Presentations can be as formal or informal as people wish. Sessions can be run as casual workshops or polished PowerPoint pitches—whatever produces the best energy, chemistry, and outcomes.

The meeting assures X-team participants that their efforts will be taken seriously. The meeting is the mechanism that focuses time, effort, and attention on the portfolios. The meeting is the end of the beginning.

8. GET THE PEOPLE

The ideal 5×5 involves twenty-five people—five teams of five people each. In practice, however, the 5×5 figure is more a guideline than an absolute. A structure of four teams of five or five teams of four also works.

At one biotech start-up in Cambridge, Massachusetts, a dynamic dozen—four X-teams of three people apiece—produced an innovation-rich list of twenty experiments. Half were pretty good; five knocked the CEO/founder's socks off. Yes, the presentations impressed. But the CEO told me his most important takeaway was that his people were much more innovative than he thought.

There's no escaping the cliché: People make the 5×5. Talent selection is the make-or-break success criterion. Personnel drive policy. A CEO or business unit leader cannot send a clearer signal than the fifteen, or twenty, or twenty-five individuals they personally pick for a 5×5. Organizations take their cue from the top. Selections speak louder than words.

Should HR or other high-profile executives pick the participants, the organization still watches and wonders "Why them?" Curiosity is contagious. Organizations understand that the act of naming X-team participants says something important about a shift in innovation culture. That's why, the vast majority of the time, X-team participants are chosen by management. Self-selected X-teams can work, but they'll invariably confront managerial headwinds. Most organizations using X-team methodology insist on picking the people they want shaping their future. They're picking winners.

Their preferred choices tend to be high-energy, ambitious individuals primed for success. Managerial prowess and technical expertise are givens. A willingness to be a team player is a must as well. Internal politics and organizational dynamics always influence the selection of talent. A majority of successful X-teams represent calibrated cross-functional selections of fast-trackers. A few more established managers are sprinkled in for gravitas. A few wild cards are thrown into the mix. The 5×5 becomes a test of collaborative character, as well as individual ingenuity.

Some organizations explicitly assign individuals to X-teams, while others encourage self-organization. Some firms pick the X-team leaders, while others ask the teams to decide how they govern themselves. The underappreciated result? The X-team methodology gives organizations important opportunities to assess the leadership, initiative, and teamwork of their talent. That's why serious HR executives like the 5×5. It enhances both their insight and their influence.

In larger organizations, selecting twenty-five or more qualified X-team participants is a straightforward, viable option. (In one case, I ran a 7×5 X-team initiative for a global company that wanted more people to play with more innovation across the enterprise.)

In smaller firms, the number of members ranges between twelve and twenty-five. Most organizations prefer more but smaller X-teams over fewer but larger ones. That is, most firms choose to field five teams of three rather than three teams of five. Top managements tend to want more portfolios, more hypotheses, and more experiments to review. They value internal rivalry. More teams mean more competition, which spurs creativity and innovation.

9. THE KICKOFF MEETING

Invitations have been accepted. The brief has been circulated. Emails have been exchanged. Portfolio and presentation parameters have been defined. The executive review team has been named—does it include the CEO and a non-executive director, or perhaps other people? The delivery date has been fixed. Now comes the kickoff.

Everyone meets. In global firms, X-team members may Skype or digi-conference in. But kickoffs held with all participants in the same place at the same time work best at boosting team effectiveness. All the X-team members are in the room to assess their mission, their colleagues, and their competitors. Questions are answered. Ambiguities are removed. The kickoff's goal is having the participants leave informed, confident, energized with ideas, and eager to collaborate with their teammates.

Kickoffs generally last a half-day. Conversations over breakfast and lunch promote comity and chemistry. Everyone is there to succeed—and all the participants know that they are being assessed.

These sessions both declare and manage expectations. The 5×5 methodology is thoroughly hashed out. Kickoff meeting owners and organizers offer model portfolios, hypotheses, and experiments for context and benchmarking. Lessons learned from successful 5×5 projects from other organizations are discussed, and so are failures. An executive review team representative outlines aspirations and expectations.

Nonfinancial resource questions are clarified. Yes, X-teams will have limited access to webmasters; no, the X-teams can't ask finance to prepare data sets; yes, the X-teams can interview call center managers; no, X-teams can't rehearse in the boardroom the night before their presentation.

The kickoff typically establishes the 5×5 road map, checkpoints, and milestones between now and D-day. Circumstances and the improvisational nature of X-team efforts virtually guarantee surprises, but deadlines should remain inviolate. The kickoff meeting illuminates the path and begins the race to the deadline.

The shorter the time window, the greater the intensity of the kickoff. If presentations are a week away, these meetings focus more on timing and tactics for high-impact delivery. But if the presentations are a month away, X-team participants invest more in awareness than urgency. In the final analysis, kickoffs should make sure that all people in the room know what they need to accomplish to make the deadline. How they do it is up to them.

10. LAUNCH

A designated 5×5 wrangler addresses questions and concerns that arise. That person tends to have a very busy week before D-day.

Checklist Questions

- Do you have a compelling rationale that works for your organization or group?

- Have you written a memo or manifesto or prepared a PowerPoint slideshow that articulates that rationale?

- Do people whom you respect find that rationale appealing?

- Have you identified the best source of support for a 5×5 initiative?

- How do you know?

- Could you and a colleague find a better source of support? Why or why not?

Rapid experimentation strips away the subsidy and luxury of the time required to observe nuance and subtlety. Sophisticated outcomes are artifacts of poor experimental design. You want your hypotheses and experiments to generate results that—like the four points of a compass—unambiguously guide the direction in which you need to go. You're pushing hard and fast for clarity and contrast. Your experiment should make your next step blatantly obvious. Design discussions and disagreements with colleagues should energize instead of paralyze. Results aren't just food for thought, but fuel for action.

Cheap experiments force simple, lean, and ingenious thinking. You literally can't afford to throw money at the problem. You're looking for the bare minimum necessary to test your team's hypothesis. Less is—and should be—more. You know that you can always spend more money. But the dedicated discipline of spending less pushes you and your colleagues to test whether short cuts lead to dead ends or pleasant surprises. The quick, rapid and convincing outcomes of your conversations give everyone a richer understanding of both the hypothesis and the experimental design.

Imperfect information becomes a feature, not a bug. The best is the enemy of the good; likewise, very good is the enemy of good enough. Seeking 90 percent certainty is a fool's errand. You're running the experiment to reduce the risks of taking the next important step. These experiments are about the craft of making "good enough" good enough, but only if you can achieve it faster and for far less. Actionable insights come in a day instead of a week; outcomes that matter take a week instead of a month. What once required a full half-year of effort and analysis can be largely uncovered in a month.

A quick tempo of imperfect experimentation outperforms slower-paced thoroughness. Forget Aesop: bet on the hare, not the tortoise. If the "mean time to discovery" can't be accelerated or compressed dramatically, you're running the wrong experiments.

Consistently capturing 75–85 percent of what you think you need to know, far faster and cheaper, is a wonderful way to navigate uncertainty and opportunity. Travel fast; travel light; travel far. Keep your eyes and your mind open. Go where the data suggest. The stronger the suggestion, the more urgently you should move.

You're generating agile momentum, and that makes innovation less risky. You're not so burdened by budgets and expectations that you can't head in a different direction quickly if that's what the data demand. But you're always moving toward achieving a meaningful result. More important, what you've learned makes sharing your hypotheses, knowledge, and aspirations with potential partners much easier. Your experiments are hooks and lures that you can use to build innovation alliances.

Will the results persuade? Will they influence? Will they move people to take action? Successful experiments are about changing minds and changing behaviors.

Coming up with five individual "blow 'em away" experiments is the path of least resistance. But it's the easiest approach for an X-team to take, which is why it's also the most common. Paths of least resistance are rarely the roads to maximum advantage. How should X-teams think through and articulate their philosophy of experimentation? Why do these experiments, and why now? What does a particular portfolio say about how a team thinks innovation opportunities should be identified and explored inside the firm?

Pushing people to take a disciplined approach to these issues turns out to be extraordinarily valuable. The implementability of the experiments forces participants to treat them less as speculative exercises than as accountable choices. Portfolio thinking pushes X-teams to come up with compelling rationales for how and why their proposed experiments are best for their organizations to explore rapidly. Portfolio thinking asks X-teammates to spend time examining the connective tissue that links the experiments that they'll champion.

The portfolio has two purposes: to insist that the X-teams look for the organizing principles that underlie their hypotheses and experiments, and to give top management a window into how individual teams collaborate to create a coherent investment perspective to present. This improves the odds that both the X-teams and their top management reviewers get more value from their interactions.

But the behavioral dynamics driving portfolio development are intriguingly diverse. Anecdotally, some teams constructively argue around the key words in their portfolio philosophy and the handful of hypotheses that they select. They'll energetically debate, for example, whether their portfolio mix should emphasize growth over profitability, or quick wins over big impacts. They're acutely aware that whatever they come up with needs to appeal to their top management. They understand that their portfolio of hypotheses and experiments are part sales pitch, part organizing principle. They want something sexy enough to command attention, yet flexible enough to allow the experiments that excite them most.

By contrast, a sizable minority of X-teams can barely control themselves. They immediately dive into crafting hot hypotheses and provocative experiments. Brimming with ideas and enthusiasm, these teams pick a coherent portfolio philosophy *after* they've picked their favorites.

But what I've typically observed, however, is that team members soon get nervous when they see the span and breadth of their brainstormed hypotheses and experiments. They quickly start looking for—or making up—links between radically different hypotheses. These teams retrospectively justify their portfolio selections and designs. That's fine. They're bonding around creating coherent rationales for their five hypotheses and experiments.

Determining which approach will lead to superior hypotheses and experiments is difficult. As interested as C-level and business unit executives are in individual experiments, they also appear to take the framing of the portfolio value statement seriously. As enterprise overseers, they're sensitive to how the teams sell the business value of their experiments portfolio.

CEOs and chief operating officers (COOs) have sharply criticized what they felt was pseudo-philosophical portfolio rhetoric— "Fortunately, your experiments seem better thought out than the portfolio you've put them in"—and complimented how well proposed experiments were packaged: "That's a clever way of bundling your experiments ..."

The most common portfolio format is simply a collection of the X-team's five best experiments. Period. Full stop. Without question, many of the X-teams presenting these portfolios had excellent experiments. However, those teams, their rivals, and—significantly—their top management reviewers felt a sense of frustration. A list is not a portfolio; a bunch of excellent experiments with no real theme or articulated connection comes off as disjointed and a little disorganized.

The whole doesn't have to be greater than the sum of its parts, but some effort to be holistic matters. Intriguingly, reviewers and team members alike observed to me and each other that a few of these excellent experiments would have seemed even better had they been an integrated part of a well-articulated portfolio value proposition.

For X-teams with sales and marketing sensibilities, no shortage of clever portfolio philosophies and positioning opportunities exist. Virtually every 5×5 engagement featured an innovative portfolio design that commanded immediate attention and respect.

My personal favorite was an X-team that Googled their company CEO's speeches and analyst conference calls—and then crafted their entire portfolio around his words. The PowerPoint slides introduced the team's hypotheses and experiments by saying that they were based on what the CEO himself had identified as the most critical strategic issues confronting the firm. Before each business hypothesis was flashed on-screen, the team displayed the relevant CEO comment from which it was derived.

Was it sycophantically manipulative? Sure. But the CEO and the nonexecutive board member in the audience both paid exceptionally close attention. The CEO's conversations concerning the strategic interpretations and implications of the team's business hypotheses were excellent. As one team member remarked in the debriefing, "I've never heard [our CEO] that engaged in explaining his views."

"Underappreciated opportunities" was another attention-grabbing philosophy. Each hypothesis and experiment centered on business opportunities that the team collectively felt were undervalued by top management. This team believed that these opportunities had more potential than the C-suite thought. The brief questioning by the reviewers about the team's definitions of what *underappreciated* and *opportunity* meant set the presentation up beautifully. Top management and the team had genuinely differing perspectives of what *underappreciated* was. This led to understandable concerns about how well the firm management communicated its priorities.

Other portfolio design philosophies are more straightforward. For instance, X-teams at one European global conglomerate selected "emerging markets" and—in an explicit nod toward rival General Electric—"boundaryless collaboration" as themes. A regional financial services firm's teams chose to stress credit card risk management and customer segmentation. Simply listening to the reasons that their people had for emphasizing those themes over others proved to reveal quite a bit about the priorities of these firms' leaders.

WHO USES THE 5×5?

The 5×5 methodology has worked well in organizations that love to play with ideas, as well as firms that, frankly, would rather draw innovation inspiration from outside consultants. Collaborative firms have welcomed 5×5s as a culturally compatible innovation approach, while top-down, quasi-autocratic leaderships have cautiously embraced 5×5s as a safe, cost-effective way to diversify their innovation spending.

Motivations matter. Know the motivations that matter most. In addition, you should know—and be honest about—your own motivations.

Do you want your organization forming 5×5 X-teams as healthy, natural extensions to the ways that you and your colleagues already collaborate to innovate? Alternatively, do you believe that the collective talent and creativity of your firm are underappreciated and require new modes of interaction to innovate?

Perhaps you feel that the 5×5 methodology isn't an optimal or even excellent way to enhance your innovation culture or process. But you've concluded that it's the most organizationally palatable option available. You're intrigued less by its "wonderfulness" than by the promise of fast, cheap, and comparatively easy ways to move your organization in what you think are the necessary directions.

Maybe you see your company's existing innovation expectations and realize that, competitively, the status quo needs a jolt. For you, 5×5s may be less about exploring a novel innovation process than presenting a constructive provocation that will add new urgency to your innovation community.

In my experience, neither C-level executives nor their direct employees have to invite or authorize 5×5 initiatives. It's never a bad thing, of course, if CEOs, CFOs or top-tier executives decide to champion a 5×5 experience. However, the most effective 5×5 X-team advocates are project leaders, senior managers, and executives who understand that their organizations need faster, more agile, and lighter-weight innovation initiatives that can produce heavyweight results. They're acutely aware of their organization's cultural, organizational, and process weaknesses around innovation results. Their frustrations make them disproportionately curious about and open to a 5×5 intervention.

Intriguingly, however, they tend not to be big-budget advocates. Their visceral reaction to business problems or challenges is *not* to ask for more money. Instead, they seek creative shortcuts and ways of making their own people more valuable. Not incidentally, they have friends and acquaintances located throughout the organization. Usually, they also have a high-level executive sponsor or mentor whose confidence they've earned.

But the most important quality that they possess is the ability to identify and communicate a sense of urgency and importance. They're persuasive, not because they're good at getting funding, but because they know how to articulate a problem or opportunity that the firm knows it has to address *now*.

THE
INNOVATOR'S
CULTURE

11 CRAFTING GREAT BUSINESS EXPERIMENTS: THREE THEMES

Experiments are remarkable media and mechanisms for exploring value creation. An ounce of simple experimentation can be worth several pounds of sophisticated analysis. Successful high-impact business experiments don't demand genius; they require ingenuity. What invites and incites ingenuity?

Mediocrity seldom evokes greatness. Unimaginative strategic visions beget uninspiring hypotheses. Typical problems typically produce typical experiments. Ordinary disagreements suggest ordinary experiments. This shouldn't surprise. In most organizations, the urgently mundane successfully competes with the strategically important.

Truly great experiments in science and technology emerge from Einsteinian intuition or Edisonian insight. Great business experiments demand a different emphasis: the discipline to ask—and answer—fundamental questions about creating economic value. Great scientific experiments don't necessarily create economic value, but all great business experiments reflect scientific values.

Genuine conflict and collaborative curiosity are fantastic reservoirs for generating testable hypotheses and compelling experiments. That's a cultural commonality of innovative science and business.

Without minimizing the importance of individual talent in any way, great 5×5 experiments overwhelmingly reflect *collaborative interaction* around three themes:

- Great strategic vision

- Great problems

- Great disagreements

"Great" is deliberately chosen. These visions, problems and disagreements aren't marginal. They're core and central to tomorrow's challenges.

Don't confuse experimentation with R&D. Significant challenges and opportunities bring out talent's greatest efforts and ingenuity. This is as true for venture-funded entrepreneurial innovators as for academic laboratories. More often than not, great experiments are products of great circumstance. That's been the 5×5 experience.

GREAT STRATEGIC VISION

Google's avowed mission is "to organize the world's information." Amazon's mission statement declares, "Our vision is to be earth's most customer-centric company; to build a place where people can come to find and discover anything they might want to buy online." The Pentagon's Defense Advanced Research Projects Agency (DARPA) states its mission as follows: "to maintain the technological superiority of the U.S. military and prevent technological surprise from harming our national security."

These strategic visions aren't achievable without a passionate commitment to bold hypothesis, experimentation, and scalable testing. The leadership and the culture recognize and reward this. Success isn't analyzed, intellectualized, or studied into existence. The vision's enormous scope means modest and mediocre experimentation can't lead to success. If people won't or don't perform great experiments, the organization itself can't be great.

As if to emphasize this sensibility, financial services innovator Capital One explicitly incorporates its employees into its corporate mission: "We take pride in encouraging our associates to think independently in a collaborative setting and to present creative ideas to senior management. It's this type of innovative thinking that is a major part of our corporate growth, as well as the overall success of Capital One."

By making "organizing information" its strategic vision, Google not only gives talented marketers and engineers permission to push innovation boundaries, it invites them to rethink what the words *organizing* and *information* should mean. The firm's successful forays into email and calendaring—initially resisted by Googlers who wanted "search" as their defining corporate core competency—were the happy byproducts of initially informal experimentation.

To reinforce its vision, Google's technical infrastructure initially lent itself to—and then was explicitly designed to support—its experimentation culture. Of course, only a fraction of experiments might initially be seen as great. But the overarching strategy provided plenty of opportunity for seemingly marginal experiments to scale into something grander.

Again, Google's initial strategic vision neither excluded nor minimized the incremental; incrementalism was encouraged as a springboard to greatness. Googlers are excited by the "global laboratory" experimentation their cloud computing infrastructures facilitate. The Google strategic vision defines a grand challenge that demands iterative ingenuity. How could the firm not end up doing great experiments that redefine the industry? The environment and expectations are there.

Amazon issued a comparable call to transformative greatness. Amazon's much-emulated recommendation engines began as a bootlegged experiment that survived several senior management efforts to kill it. What saved the nascent innovation wasn't just the fact that its experimenters were exceptionally smart and persistent—they also could make a credible case that their efforts honored Amazon's central mission. Moreover, they could make another persuasive argument: their experiment was a relatively cheap, relatively risk-free, and a remarkably rapid way to test the economic value of an important business hypothesis. As Amazon founder Jeff Bezos frequently observes, "Frugality drives a lot of our innovation."

Strategic visions inviting faster, better, and cheaper experimentation are culturally and organizationally essential. They create and affirm innovation expectations that don't equate huge, transformative ideas with huge, transformative budgets. What is the difference between "frugal" and "cheap"? The former is defined as "prudently saving or sparing; not wasteful"; the latter is described as "stingy; miserly." Those distinctions aren't subtle.

Great creativity needn't come at great cost. Great strategic visions can inspire great experimentation that neither busts budgets nor defy leadership aspirations. Just the opposite—truly great strategic visions give permission to make faster, better, and cheaper experiments great. They empower.

GREAT PROBLEMS

The surest inspiration for great business experiment is a great business problem. While strategic opportunities tempt and allure, big problems disproportionately command top management's time, attention, and respect. The Big Problem with Big Problems, however, is that they typically attract Big Solutions. That is, big problems become beacons for Big (Budget) Consultants, Thinkers and Advisors who want to launch comprehensive studies and holistic initiatives. But Bigger is seldom cost-effectively Better. Big Problems don't inherently require Big Solutions to better manage them.

Toyota, a company whose Toyota Production System (TPS) profoundly transformed the economics of the automobile industry, literally brings its own business vocabulary to problem solving. Toyota's culture has learned to strike a useful balance between its desire for process rigor and the inevitability of "on-the-line" experimentation. "Shortening the time it takes to convert customer orders into vehicle deliveries," the firm insists, is the system's strategic goal.

Much as the company successfully imported "lean thinking" to material process flows, Toyota brings "lean experimentation" sensibilities to production problems. When the company performs root-cause analysis, it doesn't want managers and workers reflexively proposing "solutions." A solution-orientation promotes dysfunctional and even destructive expectations. Toyota instead asks its people to propose "countermeasures."

Why would that be? Multiple countermeasure options may deserve experimentation and testing. The word *solution* implies that the problem is solved In real-world fact, however, the solution may ultimately create problems of its own. The countermeasure concept invites experimentation and appreciation of contingency and context. Solutions offer false promises of permanence. Countermeasures buy time and the expectation of continuous improvement. Toyota's high-quality production culture is always looking for great countermeasures. Unsurprisingly, the company reports that great problems are what inspire the most ingenious countermeasure experiments. "Great problems" inspire countermeasures that fundamentally change Toyota behavior and culture.

After doing root-cause analysis, for example, TPS innovator Shigeo Shingo devised "poka yoke" methodologies to minimize human production errors. Shingo introduced simple devices that either made it impossible to fit a part incorrectly or made it obvious when a part is missing. That prevented errors at the source—a key support of Toyota's "zero defects" process philosophy. The dynamism of just-in-time, continuous-improvement production lines demands just-in-time, continuous-improvement experimentation.

Interestingly, Shingo initially called his poka yoke technique a way of idiot-proofing Toyota's production line. In the course of discussing the benefits of this process design approach with a group of workers, he recalls, one young woman burst into tears: "I am not an idiot," she cried. A stricken Shingo immediately changed the name of his countermeasure experimentation to "mistake-proofing." Toyota takes language seriously.

From their earliest days, automobiles have posed a plethora of great problems aching for inspired innovation. Charles Kettering, an engineer and particularly gifted experimentalist, devised several inventive countermeasures that defined technology-driven innovation during the rise of General Motors (GM) to market dominance in the early twentieth century.

Arguably the most daunting interface challenge for both American automobile manufacturers *and* owners at that time was starting the car. The standard interface in 1910 was a crank. Instead of turning a key, the driver stood in front of the vehicle to grip and spin a crank, which required a lot of energy. This wasn't just a discouraging physical challenge for women and slighter men; cranking was genuinely dangerous. If the car happened to be in gear, the person could break a bone.

All manner of starter gadgets—from compressed air to springs to levers—were explored. None were reliable enough to catch on. Kettering, while an engineer at National Cash Register (NCR)—the high-tech giant of its day—had invented a small, high-torque electric motor to replace the cranks on NCR's cash registers. After two months of adaptive experimentation in 1911, his entrepreneurial venture, Dayton Engineering Laboratories (Delco), demoed its electric ignition starter, also called the Delco, to Cadillac. Cadillac bit.

Kettering's starter was introduced on the 1912 Cadillac. By 1916, virtually every American car—except Henry Ford's Model T—had abandoned the crank. Rapid, adaptive experimentation literally led to reengineering the nascent automobile industry. A great problem demanded great experiments. Kettering's Delco delivered.

GM eventually acquired Delco. Within the decade, Alfred Sloan made Kettering the company's top "big problem–solving" engineer. The automobile's popularity transformed the dominant innovation problem from the electrical to the chemical. Mass production logistics erected a process barrier that the mechanical engineers hadn't anticipated. Assembly lines succeeded in building better cars faster, but that wasn't good enough. The new barrier to production efficiencies had nothing to do with the skill of the men or the speed of the line. The insuperable production bottleneck was waiting for the paint to dry—literally.

When Henry Ford declared, "Any customer can have a car painted any color he wants, so long as it is black," the comment captured the constraints of chemistry rather than his aesthetic preference. Cars could be assembled in hours, but paint and varnish typically took over a week to dry. Black paint dried the fastest; that's why Ford used it.

But in the early 1920s, DuPont figured out how to spray faster-drying lacquers onto surfaces. Kettering seized on the chemical breakthrough and tinkered, tweaked, and experimented with it to turn spray-painting into a production-line process. By saving both time and storage space for drying cars, GM reaped enormous efficiencies. The fact that GM's customers could choose from a color palette was seen initially as a marginal innovation. Unsurprisingly, it quickly became much more than that.

Under Kettering's oversight, GM introduced its "True Blue" finish in late 1923. The new colors were a hit. More important for GM's bottom line, painting times were measured in hours instead of weeks. By 1935, heat lamps made drying a five-minute step. The bottleneck had vanished.

On the surface, these Kettering vignettes read like classic paeans to good, old-fashioned Yankee ingenuity. But that misses the point. Kettering's business brilliance rivaled his technical prowess. His success at GM wasn't the result of superior foresight or planning; it stemmed from superior rapid experimentation. Skillful repurposing of technology perfected at NCR enabled rapid—and cost-effective—experimentation at Delco won a paradigm-busting contract with Cadillac. Creating successful collaboration around chemical innovation dissolved a production bottleneck that threatened GM's growth and profitability. That this breakthrough simultaneously created a new marketing aesthetic for GM's sales was a multimillion-dollar bonus.

"Drying paint faster" or "starting cars easier" may not sound like great business problems. In reality, they demonstrably proved to be dominant industrial opportunities of their day. Not unlike Toyota, Kettering's GM succeeded not by outspending or out-analyzing its competition, but by rapid—and frugal—experimentation before scaling. This isn't—and wasn't—R&D. These were cost-effective commitments to creative countermeasure experimentation.

GREAT DISAGREEMENTS

Historically, great disagreements have inspired great experiments and even greater advances in science. Louis Pasteur's simple swan-necked glass experiments refuted Félix Pouchet's championing of spontaneous generation. Ernest Rutherford demonstrated in 1909 that J. J. Thomson's "plum pudding" atomic model couldn't be valid. An eloquent and elegant experiment can—and should—be more persuasive than the most charismatically articulate scientists. From Galileo on, successful experiments in science are as much about rhetorical power as fundamental knowledge. If they don't persuade, they don't succeed.

In theory, internal disagreements around core value assumptions can generate comparable advances in business. Differing perceptions of risk, recognition, rewards, and results understandably create arguments and disputes about what's best for the business. In practice, many firms smooth over or suppress those disagreements for understandable political and personal reasons. Potentially healthy arguments yield to nonconfrontational organizational accommodation. The leadership calculates that creative benefits to constructive conflict can't outweigh the cultural costs to cooperation. The risks are too great. The result? Data-driven decisions defer to executive intuition. Tests are run to validate—rather than challenge or explore—strategic assumptions. Managers don't craft controversial hypotheses; they produce plans to secure consensus. Experiments are what laboratories do, not businesses.

This presents a genuine schism. Many organizations have their own definitions and determinants of innovation. Deviation is heresy. Henry Ford brooked little dissent over how his eponymous cars would evolve. Sloan's GM encouraged structured rivalry—its brands could compete for desirable market segments. Toyota's emphasis on continuous improvement, market competition, and relentless production prowess gives it greater flexibility to experiment. Disagreements were simply faster and easier for the company to test. Both Toyota's culture and internal economics made experimentation a norm.

This insight is crucial: When genuine business disagreements are perceived as risky and expensive, there is every incentive to deflect and avoid them. Literally and figuratively, the perceived costs outweigh the potential benefits. Where business disagreements can be quickly, cheaply, and safely explored—*but not necessarily resolved!*—they can actually become innovation resources.

If the price is right, great business disagreements can be features, not bugs. Disagreement can be a tremendous spur to innovation and experimentation. Arguments about how the organization can create and sustain value are, arguably, the best arguments that an organization can have. The challenge—and it's enormous!—is creating business environments where participants feel confident that the benefits of disagreement will outweigh their costs. Moreover, are the arguers and advocates willing—let alone able—to design cost-effective experiments and respecting the results?

These are non-trivial challenges that strike at the dark heart of human nature and organizational politics. That said, it's easy to show that—just as for the sciences—clever, cheap, and compelling experiments can turn fundamental business disagreements into undeniably profitable outcomes.

Google offers a wealth of world-class examples of how significant disagreement can drive significant innovation. The company takes enormous pride in its data-driven innovation culture and willingness to test even cherished assumptions. User focus dominates Google's designs. Experimenting is integral to how Google explores its innovation options and surfaces previously tacit behaviors.

Marissa Mayer, then vice president of Google (she is now CEO of Yahoo), described a simple experiment that the company ran to test a very important disagreement around customer desire and satisfaction. Many Googlers—and Google users—argued that they wanted more: when people did a search, more results were more desirable than fewer. More was better.

In fact, Google's search architectures made displaying more results less of a technical challenge than originally anticipated. So Mayer asked a group of Google searchers how many results they wanted to see. Almost without exception, they wanted more than the ten results that Google normally showed.

So Google ran an experiment where the number of displayed results increased to thirty. The result? Traffic and revenue from the Google searchers in the experimental group *dropped*—by roughly 20 percent.

Google gave its users what they said they wanted, but their usage of the site plummeted. What happened? After reviewing the data, the team found an uncontrolled variable. The page with ten results took 0.4 seconds to generate, but the page with thirty results took 0.9 seconds. That half-second delay caused a 20 percent drop in traffic because it undermined user satisfaction. This need for speed transcends Google's search capabilities.

As Amazon "recommendation engine" innovator Greg Linden comments, "This conclusion may be surprising—people notice a half-second delay?—but we had a similar experience at Amazon. com. In A/B tests, we tried delaying the page in increments of 100 milliseconds and found that even very small delays would result in substantial and costly drops in revenue. Users really respond to speed," Mayer observed.

The design implications were "instantly" clear: more wasn't better; faster was. The serious disagreement between more information and faster presentation pushed the firm to err on the side of speed. As a direct result of this experimental insight, Google rolled out a new version of Google Maps that was less dense and rendered much faster—and it enjoyed substantial jumps in both traffic and usage. Rapid experiment gave rapid insight into the need for an even more rapid response.

In fairness, Google's technical infrastructure and culture—much like Amazon's—both have evolved around the core value that a great experiment is a great way to resolve a great disagreement. These are people and organizations that like to argue and "go to test" to make their point. They are atypical business cultures. MIT economist Erik Brynjolfsson describes these organizational predispositions as features of "born digital" enterprises.

That said, there should be little disagreement that disagreement can drive dynamic organizations to explore innovation opportunities that clearly matter to people. Nobody ignores experiments run to resolve or illuminate the firm's most significant or controversial business arguments.

Consequently, innovators looking to reframe organizational boundaries of need look no further than the biggest internal disagreements around value creation. Whether they are about products, services, prospects, or price, these debates guarantee that attention will be paid. Successful experiments around great disagreements require stepping back with smarter questions rather than stepping up with better arguments:

- What experiment could we do that would significantly improve the quality of the debate?

- What experiment could we run that would win the argument?

- What experiment might we perform that could create common ground for the next step?

- What's the fastest experiment that we could do to get people there?

- What useful new disagreement or argument would we want this experiment to create?

In too many firms, these questions aren't even asked, let alone answered. Leadership's role in addressing them can't be overstated or minimized. How does management listen to, lead, or facilitate the arguments that matter most? Is disagreement treated as dissent? The answers overwhelmingly shape why and whether meaningful experimentation gets done. For reasons personal, political, and cultural, organizations often find converting their great arguments into great experiments too great a challenge. The passions aroused by those arguments typically taint or overwhelm more rational behaviors. Or, more commonly, the managerial powers that be try to avoid what would be internecine warfare. At one Spanish financial services firm, for example, a seemingly "obvious" approach to serving a rapidly growing market segment was simply declared "taboo." Did this taboo represent a truly "great" internal argument the bank was having? That was unknowable. Open and authentic disagreement was effectively forbidden. The 5×5 teams knew enough not to propose either hypotheses or experiments that might surface deeper insights.

As a general rule, however, by far the most interesting and provocative 5×5 portfolios tend to be inspired by or drawn from the organization's Great Visions, Great Problems, and Great Arguments. The desire and demand to impress top management encouraged the more ambitious teams to seek innovative interpretations of "greatness" in hypotheses.

From my observation, teams adept at identifying overlaps and intersections between the three "Greats" enjoyed the greatest curiosity and engagement in their presentations. Hypotheses at the intersection of "Great Problems" and "Great Visions" or experiments addressing overlaps between "Great Arguments" and "Great Visions" proved remarkably energizing.

The chance to combine a "Great Vision" opportunity with an approach to solving a "Great Problem" was irresistibly appealing. The ability to connect a "Great Argument" to expand or extend a "Great Vision" excited top management.

Getting 5×5 teams to explicitly list and then map "The Three Greats" virtually guarantees that strategically important hypotheses will emerge.

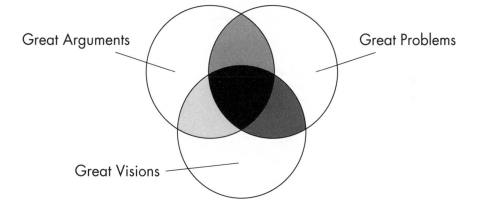

12 A GUIDE FOR X-TEAMS

Creating high-performance X-teams is readily achievable. Talented people—representing a mix of organizations from all over the world—have achieved remarkable results remarkably fast. Their teams did great. Their experiments impressed.

This chapter explores and explains those critical X-team success factors. Respecting simple and fundamental X-team principles shifts the likelihood of success from possible to probable.

If you're reasonably committed to being a good X-team player, you will get good results for yourself, your team, and your organization. If you and your colleagues are seriously committed, you will get outstanding results. The purpose is better aligning your team's internal actions with your desired outcomes.

The most important questions facing your team: What do you want the outcomes to be? What impact and influence does your team want to have on top management and your organization? How should your portfolio of hypotheses and experiments change minds or influence perceptions? What team experience do you want to have and share? What results will create both satisfaction and success?

Your team will fight, bond, or achieve an uneasy truce around these questions. But authentic individual and collective efforts to answer those questions in a serious way overwhelmingly determines your team's success. Why? Because that struggle will reveal the talents, concerns, ambitions, diversity of views, and ingenuity that are essential to crafting business experiments that command attention and respect. Serious X-teams get real results.

The good news: Almost all successful X-teams have serious arguments around business themes and hypotheses as they design their experiment portfolio. The bad news: So do underperformers.

There are unhappy and argumentative X-teams that produce brilliant experiments, and there are happily collaborative X-teams that generate mediocrity. Conversely, there are unhappy failures and healthy collaborations that produce remarkable portfolios. Neither comity nor chemistry predicts productive consensus or unfortunate outcomes. How well X-team members like—or dislike—each other doesn't appear to determine work quality.

Nevertheless, the next most important question facing your team is: Who is on your X-team, and are they committed to success? As a rule, the overwhelming majority of X-team participants represent the firm's most talented and competent people. Getting them to cohere into collaboratively productive—and productively collaborative—X-teams is typically more a challenge to be met than a problem to be solved.

The discipline required to designing and delivering an innovative presentation to top management on a demanding deadline tends to minimize interpersonal foolishness. That said, there's no shortage of excellent research and literature on enhancing small group performance. X-teams offer the talented, ambitious, and energetic a fantastic opportunity to show how well they work with others.

Five critical success factors stand out. They are remarkably consistent. They transcend geography, industry, and business culture.

Every X-team regarded as successful—in the moment and in retrospect—displayed these attributes.

High-Energy Focus on Potential Business Impact

Don't think small. Successfully regarded X-team portfolios explicitly addressed *Big* issues and opportunities. No subtlety or nuance here. The teams emphasized experiments that could have a clear and major impact on customer or supplier perceptions, or on how the firm saw itself. Hypotheses and experiments aggressively addressed the left side—not the right—of the organizational decimal point.

X-teams presented stories, narratives, and vignettes emphasizing *the* future strategic or operational import of the proposed experiments. Seemingly smaller or more tactical experiments ultimately were linked to bigger and broader business themes. Proposals reflected strategic leadership issues more than management concerns.

Successful X-team presenters made the case they had discovered novel, quick, and cost-effective ways to help top management test high-impact opportunities. These weren't marginal business value issues. Top management generally agreed these X-team themes and associated experiments deserved their time, attention, and investment.

Clear Alignment with Known Top Management Priorities

Respect the boss. Successful X-team portfolios explicitly acknowledged the strategic business goals that came from the organization's C-suite. While these teams possessed their own strategic sensibilities, they fully grasped top management's priorities as well. Even out-of-the-box portfolios and experiments were presented in the context of top management's strategic intent.

It's important to note that several X-teams did offer up contrarian—even defiant—portfolios and experiments that were (in my opinion) excellent and constructively provocative. However, some of them did this in a manner top managers experienced as disrespectful or insubordinate. In other words, good portfolio, bad reception. The importance of respectful dissent and challenge cannot be overstated. Poor presentational positioning undermined and overwhelmed the value of these X-teams' insights.

Successfully aligning—or linking—portfolios to top management priorities facilitated healthy conversational give and take. Leadership's questions and comments revolved around how X-team hypotheses and experiments could help the firm manage business outcomes better. Presentations became springboards for top management discussions that never would have otherwise occurred.

Simple, Compelling, and Accessible Presentations

Don't be difficult. Top managements liked proposals that feature simplicity, accessibility, and usability. Successful X-team portfolios were never exercises in complexity or detail. They were easy to follow, easy to understand, easy to believe, and easy to share. They possessed obviousness. Top executives couldn't help asking, "Why hadn't we thought of this in that way before?

Extraneous complications and technical jargon had been stripped away. What remained were clear hypotheses and intuitively appealing experiments. Graduate degrees, external consultants, or doctoral students weren't required. Simplicity makes clarity easier, and greater clarity heightens awareness and involvement.

One X-team's presentation crashed and burned when the group struggled to explain a complicated and confused experimental design. It was embarrassing. The core idea was quite good; its articulation was awful. More thought had gone into refining the details than in communicating its strategic value. The group's mismanagement hurt the rest of its portfolio.

The "easy to follow/easy to get/easy to follow up" design of X-team portfolios increased audience confidence that people would be inspired and energized. Don't make top management work too hard to appreciate the brilliance of the hypotheses and proposed experiment.

Pleasant Surprises

Pleasant surprises win. Successful X-team portfolios feature pleasant surprises. Sometimes they are embedded in business value hypotheses that hadn't been previously considered. For example, one X-team surprised top management by hypothesizing that key customers might get significant value from using a diagnostic tool already being actively utilized within the firm. Repurposing these successful internal tools for external use never crossed senior management's minds.

Surprise also comes from experiments more revealing than status quo analytics. One firm found purchasing Google's AdWords could identify potential new product-testing partners far faster than its existing qualification process, which took 90 days or even longer. Back-of-the-envelope calculations suggested the firm could ascertain 80 percent of its desired qualifying objectives in two weeks. This sparked a firmwide rethink of its beta testing and go-to-market processes.

Top managements are all too familiar with unpleasant surprises; they prefer pleasant ones. Successful X-teams understand the appeal of desirable novelty. Hypotheses or experiments that simply validate or affirm top management's business concerns don't excite. "Design for surprise" boosts executive interest and involvement. X-teams unable to surprise themselves in a good way are unlikely to surprise top management that way, either.

Scalable Next Steps

No scale, no sale. Successful X-teams look beyond the immediate experiment. They anticipate what's next. Like chess Grandmasters, these teams have clearly thought two to three steps ahead. They've pondered how the firm could scale their small experiments into bigger business projects, programs, or processes. Although they haven't solved the problem—they're not supposed to!—they've considered possible options.

Holistic thinking matters. Successful X-teams know that their presentations are more about beginning an innovation journey than finishing an experimentation exercise. Presenting a persuasively provocative portfolio isn't enough. If the portfolio is implemented, what would happen? How do individual experiments suggest their even more innovative successors? How should heavyweight hypotheses become platforms for innovation? Smart X-teams have plausible answers for these questions.

Anticipatory awareness impresses top management. Successful X-teams can't map the future, but their portfolios are positioned to help point their organizations toward it. They know how to communicate where their firms could go next. Successful X-teams bring true top management sensibilities to their top management presentations.

X-teams that can't deliver at least three of these five elements discussed here seriously jeopardize their shot at success. Understandably, legitimate disagreements around surprise, simplicity, and scalability exist. But successful X-teams push hard to ensure that their portfolios are designed to impress. They'll test-market their hypotheses and experiments with people both inside the enterprise and out. They're constantly seeking constructive feedback. They're looking for an edge. They want their portfolios to have symmetric left-brain *and* right-brain appeal: rational experiments that inspire and passionate hypotheses that provoke new thinking.

THE X-TEAM

X-teams typically have five people that represent a cross-functional slice of the organization. They're there because management believes that they're innovative, talented, energetic, and destined for greater things inside the firm. They want X-teams to succeed; they want to be impressed. They seek confirmation of their own acuity. They believe that 5×5 initiatives is a terrific innovation investment and that they've selected their best people to make it work.

Your team's job is not only to impress your organization's leadership, but also to produce a portfolio that impresses you. You want to surprise yourself pleasantly; and you want your colleagues to surprise each other (and you) the same way. That happens by building on each other's insights and expertise.

X-teams present opportunities for their members to have a positive strategic impact on your organization, individually and collectively. The team's energy, diversity, and talent should be a creative source of move-the-needle innovation. You are looking to your colleagues—and they to you—for insights and arguments to build experiment portfolios that capture the imagination and spark enthusiasm.

Your team fails if your conversations become echo chambers for top-management clichés. Your team fails if its business insights sound like escapees from tired executive education webinars. Your team fails if your business hypotheses read like last year's strategic visions. Your team fails if your proposed experiments sound like warmed-over market research. And *you* fail if your portfolio doesn't reflect your team's intrinsic ingenuity.

Success begins with respecting your team's talents, your firm's needs, and your market's potential. Success comes from self-organizing to tap individual and collective experience and expertise. You can appoint or elect a leader. You can go the participatory democracy route. Structure your interpersonal interactions any way you please. How your X-team manages its meetings is up to you.

The central question: Can your X-team design itself so it can deliver on each of the five critical success factors? Are you, individually and collectively, prepared to present a portfolio that deserves leadership's undivided attention? Does your team accept that crafting hypotheses and designing experiments demand collaborative discipline that goes beyond celebrating good ideas?

Actions speak louder than words. The 5×5 X-team process requires specific actions to produce "shock and awe" portfolios. These actions aren't sophisticated or complicated. But they require collaborative commitment. An unwillingness or inability to work together guarantees underperformance.

Two particular 5×5 X-team dysfunctions merit special attention. These dysfunctions aren't common, but they appear often enough to warrant trying to circumvent them:

1. *X-team members who, in fact, are not team members.* Ego issues, personal agendas, or pet business themes define these people's interpersonal behavior. They appear less interested in participating in design conversations than in dominating them. They see the X-team as a vehicle to advance—rather than examine and develop—their own ideas and ambitions. They appear to listen to others, but only with an ill-concealed intention of bringing the discussion back to themselves. They're the stars—the rest of the team is the supporting cast. And that just doesn't work.

Some of these people are irredeemably manipulative jerks (whose inclusion on the X-team reflects a failure of executive judgment). More typically these are bright, passionate overenthusiasts who can't understand why their X-teammates don't appreciate their brilliance.

Yes, they have interpersonal issues, but these might be surmountable. If teams can convert socially awkward overenthusiasts into quasi-functioning colleagues, their benefits can outweigh their cost. This truly tests a team's overall character and leadership.

2. *Treating X-teams as a collection of talented individuals rather than a collaborative team.* Some X-teams fall into the easy trap of having people go off on their own to develop their hypotheses and experiments. Everyone will supposedly review each other's work and select the best individual proposals.

This system invites dysfunction, though. Absent a serious attempt to forge common understandings, people go away with different perceptions of the task. When everyone next meets, the discussions invariably revolve around those differences. Conversations begin with disagreements around those differences, and it takes time to sort all that out—time that you could have been spending on the actual issue or problem.

The "holistic team" ideal gives way to individual advocacy. People champion their pet hypotheses and favorite ideas. Of course, several proposed hypotheses and experiments are excellent. But everyone knows that they're the product of individual ingenuity and creative collaboration. That's both a strength and a weakness. Shared development and joint ownership is hard. These X-teams typically don't offer portfolios; they present lists.

To my knowledge, no X-team going the "smart individuals" route has been thought a top presenter by their top management. The quality of an individual experiment apparently doesn't outweigh the absence of a high-performance team portfolio. Top managements want both excellent experiments and excellent teams. Can X-teams succeed by listing their very best individual hypotheses and experiments? Perhaps—but I have never seen it done.

The most successful X-teams—those deriving the greatest sense of accomplishment from the process—stress the collaborative value of interpersonal interaction. "We couldn't have done this on our own," is their refrain. The whole genuinely seems greater than the sum of the parts. If it isn't, the X-team has underperformed. People can see and sense that.

SUCCESS BEHAVIORS

Begin at the End

Don't begin at the beginning. The deliverable is a presentation featuring an innovative portfolio of five business hypotheses and experiments. That deliverable is the means to your team's desired end. Start there.

Begin with discussions about the impact that your team wants its portfolio and presentation to have on management. Those discussions sensitize everyone to the business themes that matter most in achieving that impact.

Your presentation may last anywhere from twenty-five to fifty minutes. There will be questions, interruptions, and follow-ups. What conversations do you want this presentation to provoke, influence, or inspire? Why? Are those the most important conversations that your management—and your organization—need to have?

Identify that conversational impact. Reverse-engineer the presentation that your team needs to make to achieve it. If your presentation—your hypotheses and experiments—stimulates high-impact conversations that can change minds, you're successful. If it inspires executive action, you've been remarkably successful.

What's the Story?

Isaac Newton had his apple. Galileo Galilei had his Leaning Tower of Pisa. James Watt had his whistling teakettle. Apocryphal or not, these tales of experimental insight memorably capture mind share. Simple experiments can reveal powerful fundamental truths.

What is the story behind your proposed experiments? What stories should your experiments tell about your business, your customers, or your opportunities?

Behind every good hypothesis is a good story. Behind every great experiment is a great story. What's yours? What makes it good? If it's only good, what might make it great?

Effective rhetoric is the power to persuade. Compelling stories that explain experiments and simple experiments that tell compelling stories define rhetorical effectiveness. A hypothesis is also a story. Experiment helps determine whether that story is closer to fact or fiction.

Insist that every hypothesis explored and experiment proposed become a story that your top management wants to share with others. One X-team told management, "Our biggest customers would like us more, think we're super-smart, and save us money if we gave them one of our best diagnostic tools before they gave us their custom designs to build."

This simple story of sharing internal tools with external customers ultimately became an important design sensibility for the firm. It inspired a new genre of innovation. Internal tool builders began seriously thinking about which people outside the firm could get real value from their investments.

Your Experiments Should Create Heroes

Stories, hypotheses, and experiments don't need heroes to succeed. But successful leaders and managers create opportunities for their people to shine. X-teams should exploit that. Business experiments don't just test hypotheses—they give people a chance to become "innovation heroes." In other words, X-team experiments provide platforms for colleagues to look good. The ability to make people look good can be powerful.

Do your experiments require internal webmasters, call-center managers, sales teams, or customer service representatives to add value? Can their performance be a source of celebration for the firm? Top managements like initiatives that give employees the opportunity to impress. If your experiments create opportunities to raise the participants' profiles, that invites energy, engagement and support.

Stone Soup Delights Executive Palates

Convince top management to see the collective contribution. Everyone knows the "stone soup" story. That's the tale of a poor but clever cook who begins boiling a broth featuring a single stone in a pot placed at the center of town. But he persuades his fellow townspeople to contribute bits of meat, vegetables, spices, and other ingredients to make his stone soup tastier. Of course, his soup would be appallingly bland without the community's culinary participation.

Similarly, hypotheses and experiments will appear richer and tastier if executives can sample the flavors of individual contributions by X-team members. Seeing how research and development influenced a marketing experiment, or how legal counsel shaped a behavioral economics pricing hypothesis, spices up portfolio stories. Top management gets a greater awareness of how people successfully hurdle functional boundaries to create new value. They see how individual talents blend into creative collaboration.

Make Your Bosses Feel Smart(er)

X-teams present to impress top management, but C-suite executives often want to impress the presenting X-teams. Sometimes they want to be the smartest people in the room. Top managers rightly believe that they can add value to even the best ideas and proposals. Let them. Encourage them. Invite them.

As polished and professional as your portfolio presentation may be, treat it as an invitation for your bosses to contribute. Yes, it feels wonderful if the CEO or chief marketing officer wouldn't change a thing about a proposed experiment. But having your boss suggest constructive improvements to your portfolio is wonderful, too. Greater involvement indicates greater interest.

At one X-team's presentation, the CEO leaned forward after one provocative marketing experiment had been proposed and said, "I think I know what you are trying to do here, but wouldn't this experiment be better if you focused on X ...?" In fact, the CEO's suggestion would make the team's proposed experiment notably better. This was a win/win/win situation: the X-team was genuinely impressed by their CEO's insight; the CEO had the opportunity to build meaningfully on the work of a talented team; and everyone in the room felt smarter for the interaction. Healthy interactions build morale. Morale is an asset.

Executive interjection that tweaks or fine-tunes an experiment or hypothesis is not the issue. Neither are alpha-male displays of C-level dominance. Luring your bosses into revealing the best of themselves is the goal. A CEO should display the judgment, experience, and strategic emphasis that an X-team is unlikely to possess. Both the executive and the team should be pleased with the contribution—even if it is a criticism.

Good leadership can look smart without making its followers feel stupid. Give them the chance to do so.

Heed the Need for Speed

Faster really is better—much better. Carrie Fisher, the novelist/actress best known as Princess Leia in the *Star Wars* movies, once wrote, "Instant gratification takes too long." Any trick, technique, or technology that your team uses to run an experiment in a week instead of a month—or a day instead of a week—is a huge win. Go for it.

Even if back-of-the-envelope calculations indicate that running an experiment over three weeks would give you results one-third better than quicker and dirtier ones that you could run in a week, always lead with speed. You can unveil the more rigorous experiment later. But when chief financial officers (CFOs) or top sales executives say, "You mean you can actually run this experiment and give me usable data to decide whether we should go ahead with a pilot program by this time next week?" your X-team's ability to answer "Yes" and mean it, credibly, can't help but impress.

Speed exerts a seductive discipline, dynamic, and focus for top executives. Make sure that you have a couple of portfolio experiments that can be run in closer to five days than five weeks. If you propose an experiment that can genuinely move the needle in five hours or less, your X-team will be stars. Design experiments that deliver strategic insight in five minutes, and your X-team will become legends. How about five seconds? Why not?

Craft No-Lose and Win/Win Hypotheses

Avoid failed experiments. No single piece of innovation advice is more wildly—or more fatally—misunderstood than "Fail faster, succeed sooner." Attributed to brilliant industrial designer and IDEO founder David Kelley, the sentiment is laudable, but its reality is dangerously inappropriate for most organizational cultures.

The underlying belief is that learning from failure rapidly leads to successful outcomes faster. This recalls Thomas Alva Edison's famous quote that he didn't have a thousand failures on his way to inventing the lightbulb; he had actually discovered a thousand ways that didn't work.

This sounds inspirational and motivational. Don't be fooled. Does your organization honestly have the time, resources, and intestinal fortitude to endure a dozen failures—let alone a thousand? Does your X-team? Do you? Honestly?

Avoid the foolish—and politically correct—trap that failure is just another step on the path to ultimate success. It's not. Anticipating or welcoming failure is exactly the wrong way to approach business experiment design. Even Edison—a master marketer as well as inventor—appreciated that *failure* was a bad word. Business hypotheses should be designed to give the best of both worlds: if the hypothesis proves valid, the experiment gives vital insight into value creation. But if the experimental results don't support the hypothesis, the experiment still gives vital insights into value. That defines the no-lose, win/win hypothesis.

The no-lose hypothesis is the innovative way to go. Crafting win/win hypotheses requires authentic ingenuity and a sure grasp of the important. This is (arguably) the most difficult design competence for X-teams to master. Coming up with a hypothesis where top management will be fascinated by literally any outcome is the true test of an X-team's talent.

A journalistic example offers a simple framework for thinking about this business hypothesis design. Picture a horrendous, horrible, and destructive tragedy striking a major urban area. A story about how a comparable city was taking steps to prevent such an occurrence from ever happening would surely be on page one of the newspaper. Conversely, a story about that city having the chance to prevent such a catastrophe—but doing nothing—also would deserve front-page placement.

In other words, a well-reported story about taking action or *not* taking action merits page-one coverage either way. For ambitious journalists, this is a win/win, no-lose investment. The journalist, the publication, and its readers win either way.

Bring this spirit and substance to hypothesis design. If the business hypothesis turns out to be correct, top management and the firm learn something important. If not, top management and the firm learn something important anyway. Win/win. No-lose. Happy result.

A real-world X-team example about Internet travel illustrates this point. An online agency was looking for ways to capture marginal additional revenue from airline seat maps and seat selection choices for passengers. An X-team proposed a hypothesis to test whether passengers—particularly families with children—would pay a small premium to be able to sit together on the plane. That is, would a family of four or five be willing to pay an additional $6 or $10 per seat to be able to sit together along a single row or back to back?

This no-lose hypothesis turned out to be particularly interesting and provocative. If travelers were willing to pay extra to sit together, that could mean millions of dollars of additional profit for the firm and its airlines. If not, what did this reveal about opportunities to brand "family flying" as a market niche or segment?

Seeing which families might pay a premium to sit together offers win/win insight into huge slices of the firm's "leisure travel" business. Either way, this experiment generated actionable information about an important customer segment. Top management was genuinely interested in—and concerned about—either outcome.

Of course, not all portfolio hypotheses can be win/win or no-lose. Some experiments have X-team members—and top management—rooting for particular outcomes. That's inevitable. However, X-teams should make every reasonable effort to go beyond hypotheses that smell of "desirable vs. undesirable" or "good vs. bad" results. Whatever teams can do to structure hypotheses that encourage more people to get more value from either outcome is worth it. By definition, well-designed win/wins can't lose.

If You Can't Afford to Have It Fail, It's Not Really an Experiment

Don't cheat yourself. Cheapness is a beautiful element of the 5×5 methodology. In theory, the cost of running the experiment is so low that it's economically irrelevant. In practice, many costs go beyond the budget: cultural, organizational, interpersonal, managerial, social, and psychological costs matter, too.

The 5×5's key purpose is reducing real costs and perceived risks of exploring high-impact experiments. Many individuals and organizations, however, discover they feel professionally and emotionally invested in their hypothesis. They desperately—almost pathologically—want their hypothesis to prove true. They want their experiment to be successful. It's akin to flipping a coin but passionately praying for heads. When tails turns up, the instant visceral reaction is "two out of three ..."

Organizations that find themselves spending tens—and then hundreds—of thousands of dollars on experiments lose their objectivity. They've eliminated their ability to be neutral about underachieving results. Investing so much time, effort, and resources into a so-called experiment means that they literally can't afford to have it fail. Failure is simply too painful. So they start tweaking or fudging the outcomes. They want the results to look a bit better than they really are. They cheat.

The same pathology occurs with hypotheses and experiments that X-teams care too passionately about. They've fallen in love and behave accordingly. They honestly believe a particular hypothesis, experiment, or both are profoundly transformative. Failure is not an option.

Don't indulge that sentiment. Passion is super-important for innovation success. But dispassion is vital for assuring integrity and credibility. If your X-team can't afford to have its business hypothesis fail in front of top management, it's no longer a hypothesis—it's a cause. Don't be blind.

Surveys Aren't Experiments; Experiments Aren't Surveys

Market research questionnaires aren't hypotheses. Asking people what they want isn't an experiment. Having individuals rank the relative desirability of imagined features and functionality isn't an experiment either.

The most pervasive design flaw perpetrated by X-teams is declaring a set of questions is an experiment. It's not.

Questions and questionnaires can be important parts of experimental design. A completely legitimate experiment, for example, might hypothesize how men versus women or singletons versus married individuals might respond to particular questions.

But the central unit of analysis for experimental design is not the question, but rather the offer. For example, the flying experiment discussed previously didn't survey the individual willingness of flyers to pay extra to sit together. No web-based questionnaire asking people if—given the opportunity—they might be willing to pay $5, $10, or $20 more per ticket to sit together was used. Rather, when travelers indicated that they were buying more than two tickets, a question popped up on the screen: "Would you like to sit together?" If—and only if—the answer was "yes" would the site switch the customer to the experiment's screen. That's where contiguous seating for a small additional fee was tested.

Business experiments are best situated in the behavioral response to offers than intellectual answers to questions. Of course, experiments don't have to involve customer choices at all. Reviewing 5×5 experiences indicates that a noticeable portion of X-teams fall into the rut of believing that surveying people—whether customers, partners, or prospects—is equivalent to devising a hypothesis conducting an experiment. It's not.

Make Sure Your Hypothesis Demands an Experiment

Is your 5×5 experiment the most compelling way to test your hypothesis? Nothing kills an X-team's credibility faster than superior competition. If a quick review of existing data or a call to in-house experts can offer greater insight faster than the suggested experiment, then don't experiment. Top managements are rightly unimpressed by experiments that only replicate or marginally extend existing knowledge.

The 5×5's underlying rationale is not experimentation for its own sake, but persuasive evidence that good experiment trumps good analysis. If a Microsoft Excel spreadsheet or email to Joe (who's been our smartest customer for 20 years) offers more useful information at less cost than a brilliant 5×5 experiment, why do the experiment?

Yes, experiments can complement or enhance analysis and advice. More important, well-designed experiments play vital roles shattering conventional wisdoms, false truisms, and flawed assumptions. But the reason that most X-teams and top managements should be enthusiastic about 5×5 portfolios is value for time and money. These experiments offer the most economically useful way to produce invaluable insight.

You run the experiments precisely because there's no "Joe" or spreadsheet model that meaningfully gets you from here to there. You run the experiments because they'll offer results that push organizations to talk with Joe or to build models to innovate better. You do the experiment because the X-team believes that expert and analytic limits have been reached.

The experiment is an inarguably essential ingredient in the innovation mix. Don't take that for granted. Your organization may be smarter than your X-team credits. An exhilarating—and disheartening—fact that X-teams frequently discover is that their organizations possess enormous quantities of tacit knowledge. The right question, the right challenge, or the right hypothesis is the lure that helps make tacit knowledge explicit.

Consequently, X-teams become important organizational vehicles not just for experimental design, but shared enterprise knowledge and expertise. Conversations around experimental design are invariably discussions around knowledge, expertise, and analysis.

Any experiment proposed by the X-team that disregards or lacks accessible organizational knowledge can be potentially devastating. The X-team ends up looking lazy or incompetent—or both.

Don't Dismiss the Devils of Detail

Simple experiments aren't always easy. Easy experiments aren't always simple. The gap between your X-team presentation and what can be implemented quickly can be deceptively difficult. You look foolish if experiments that you declare will take three weeks to run takes three months just to set up. The courage and credibility to scope proposed experiments is imperative. Crossing every *t* and dotting each *i* isn't necessary; not knowing the alphabet, however, is unacceptable.

If an experiment requires developing a simple Facebook application, make sure that you've got an internal information technology (IT) developer (or maybe just an average teenager) capable of producing one. The same holds true for iPhone or Android apps. Likewise, if an experiment requires paper coupons placed in supermarket baskets, make sure that you've got quotes from a copy shop and email addresses for some store managers or district supervisors.

The 5×5 methodology doesn't require a comprehensive discussion of portfolio implementation. But top managements and X-teams should be confident that their proposals are readily doable. The guiding principle should be that if top management gives the go-ahead to deliver the proposed 5×5 portfolio ASAP, the X-team's reaction will be a delighted enthusiasm.

Admittedly, it's impossible to know in advance just how hard it is to turn simple and straightforward designs into real-world experiments. But showing that the X-team has scoped the first 48 hours of implementation is a wonderful discipline and confidence builder.

That exercise unfailingly proves to be revealing. X-team members gain keener insight not just into the mechanics of an experiment, but the underlying themes of the broader hypothesis. The act of implementation is always an act of discovery. That's a virtue.

Don't Be Excellent; Be Third

"Good enough" isn't always good enough. But pursuing experimental excellence often paralyzes more than it stimulates. Avoid diminishing returns. X-teams should strive for the fastest, simplest, cheapest, easiest, most accessible, most reliable, and most persuasive ways to test their hypotheses.

Given the constraints, your experiments always will be flawed. That's why "good enough" isn't good enough. Clear and convincing evidence must emerge from your experiments. Turn those flaws into strengths. Use them to invite constructive scrutiny. Your team's challenge is designing experiments that produce "clear and convincing" as quickly, cheaply, simply, and transparently as possible. Defining "clear and convincing" is hard, but it sits at the core of your effectiveness.

Sir Robert Watson-Watt, the diminutive Scotsman who pioneered radar's rapid deployment in Britain during World War II, was a master of experimental demonstration. His ability to persuade his funders quickly, cheaply, and convincingly was essential to the development of the Chain Home network of radar stations that helped win the Battle of Britain. His "Culture of the Imperfect" essay beautifully captures the design ethos that made his investments in experiment so successful: "Give them the third best to go on with; the second best comes too late; the best never comes." Strive for third best.

You Have to Want It

Talk is cheap. Yes, the 5×5 methodology is an intellectual, interpersonal, and presentational exercise. But that misses a key point. Nothing confers greater credibility upon an X-team than a confident willingness to implement its idea. Competent top management can tell whether X-team presentations are simply going through the motions. These people can tell when X-teams have become truly committed to their portfolios. Passion matters. An X-team's ability to inspire confidence is part of its leadership challenge.

Would your X-team want to run the experiments that you've designed? Would you stake your professional reputation on the experiments that you propose? Would you bet your bonus or your annual review ranking on their import and impact? Trust me—people can tell how committed you are.

I've seen X-team presentations filled with fascinating hypotheses and truly provocative experiments. The discussions around them were stimulating. But everyone in the room could tell that these presentations were where these experiments went to die.

The X-teams put on an excellent show. But this is where the show ends. These organizations embody the reality that innovation isn't about the quality of ideas, but the courage and conviction to implement.

Conversely, I've seen X-team presentations with portfolios that outsiders might view as obvious or mildly clever. However, these teams and top managements were clearly energized. They wanted to take the next steps. They wanted to innovate. But they feared the costs, risks, and complexities. These X-teams experiment proposals boosted confidence in their people and in their innovation investments.

The bottom line: Treating 5×5s as intellectual exercises typically becomes a horribly self-fulfilling prophecy. Go beyond intellect.

X-teams and top management who see 5×5 presentations as the start of healthier and more agile innovation cultures behave differently. They understand that successful innovation goes beyond plans and planning.

Successful innovation requires a commitment to action. As Joseph Schumpeter remarked, "Innovation is less an act of intellect than an act of will." Wanting to turn hypotheses into experiments into innovations is an act of will. Successful experiments invite commitment.

13 WHAT MAKES "HYPOTHESIS" SO HARD?

During a 5×5 workshop in Sydney, the team leader at an elite professional services firm proudly presented her group's portfolio of business hypotheses and experiments. Alas, it contained neither hypotheses nor experiments. Instead, the 5×5 featured detailed proposals and ambitious plans. Nothing was fast, cheap, or simple. The proposals sounded remarkably like what elite professional services firms pitch to win clients.

When this was politely pointed out, the response was aggressively defensive: "We thought that by taking a little more time and investing more resources, we could come up with actual solutions to problems instead of incomplete experiments. We wanted to do it 'right.'"

Besides, the team insisted, aren't proposals a kind of hypothesis? We predict we can reap the benefits of our plan if we execute it well. The plan is the experiment. Our organization will learn as we implement. Q.E.D.

This extreme (but true) vignette is not atypical. It highlights the singular challenge to organizations seeking real value from experiments. Creating simple, compelling, and readily testable business hypotheses is managerially unfamiliar, uncomfortable, and unrewarding. So managers avoid them.

Smart, knowledgeable, and well-intentioned executives find disciplined thinking around business hypotheses difficult. That's a global phenomenon. My experiences suggest organizations prefer their problems and opportunities be defined in terms of requirements and plans. Managerial mindsets typically value comprehensiveness over simplicity and plans over tests. Good answers make more managers happier than good questions.

Behavioral economics research also suggests managers tend to be creatures of confirmation bias and wishful thinking. Most teams and organizations, for example, consistently underestimate how much time, money, and effort their plans will consume. They're overoptimistic to a fault. This "Planning Fallacy," a term coined in 1979 by future Nobel economist Daniel Kahneman and collaborator Amos Tversky, pervades personal and professional life. Understandably, organizations committed to their plans want them to work; they give preferential treatment to evidence that affirms their success.

Confirmation bias consequently reinforces or even exacerbates errors of optimism. The more comprehensive the plans, the more susceptible the planners become to these pathologies. These behaviors aren't aberrations; they're global phenomena. Confronting them is essential to making 5x5s work.

The pressure of constantly having to handle everyday events gives organizations a plausible excuse not to bother with experimenting, even if they're technically capable of doing it. The inertia of existing perceptions, norms, and incentives undermines executives' willingness to rethink how they think. The talent and creativity may be there, the innovation imperative is not. Culture conquers cognition.

Put another way, leaderships aren't too lazy to explore the business case for hypothesis. They're just too busy trying to do their jobs to take time to test their innovation assumptions. They're concerned that experiments may be riskier and more time-consuming than they're worth. In their minds, doing an experiment is likely to be more expensive than not doing an experiment. So why bother coming up with a hypothesis? It's a waste of time and money.

But time and money are rarely the real issues. Indeed, minimizing time and money is central to the 5×5 methodology. Ironically, crafting compelling hypotheses and simple experiments is more about new patterns of behavior than cognitive limitations or costs.

As Marvin Bower, who in effect founded McKinsey & Co., incisively observed, culture is "the way we do things around here." Hypothesis is simply not part of the way most organizations do things. Organizational vocabularies and expectations surrounding hypothesis generation and experimental design are limited. Most firms have internally "branded" their planning and analysis process as epistemologically superior to hypothesis and experimentation. Organizations behave as if simple experiments are incidental—rather than central—to actionable innovation insight. That's the core cultural issue.

Literally, the opposite sensibility defines elite scientific and technological innovation. The rhetorical and substantive importance of a provocative hypothesis to real-world innovators is difficult to overstate. The Googles, Amazons, Apples, Netflixes, and Capital Ones—not to mention world-class research universities—fluently hypothesize and culturally commit to experimentation in ways utterly alien to most commercial enterprises. These firms don't insist on performing lots of interesting experiments because they're rich; they're rich because they insist on performing lots of interesting experiments.

These experiments aren't improvisational exercises in "throwing something against the wall to see if it sticks." To the contrary, experimentation here is rooted in rigor. The hypotheses specifically reflect and respect the organization's concerns—or, perhaps more important, the concerns of their entrepreneurial founders.

These conventional circumstances help explain why organizations pathologically overinvest time and resources in more traditional planning and data analysis. Unfortunately, managerial investment in business hypothesis and simple real-world experimental designs is both marginal and marginalized.

Most executives, for example, can capably describe their firm's top strategic initiatives. But how comfortable or confident would they be detailing the business hypotheses underlying them? That language is typically absent from everyday vocabularies.

Incentives also matter. Consider how many businesses internally recognize and reward new ideas. Now—how many companies explicitly celebrate and share provocative hypotheses and experiments? Those celebrations are cultural and operational norms at Capital One, Amazon, and the like. Barely a handful of the companies I've advised formally reviewed or funded business experiments before the 5×5. Projects, proposals, pilots, programs, and plans, yes; hypotheses and experiments, no. As management guru Peter Drucker aptly observed, "What get measured gets managed." Who measures the business hypotheses that matter?

This institutional dismissiveness has been assiduously cultivated in the world's most prestigious business schools and executive education programs. For example, while most of the top-twenty business schools offer vibrant business plan competitions, what programs, prizes, or classes celebrate artfully crafted business hypotheses or experiments? The overwhelming majority of elite MBAs never were educated, trained, or encouraged to design business experiments.

In effect, traditional business cultures worldwide privilege planning and discriminate against hypothesis. Why shouldn't MBAs and C-suites behave accordingly? In the "choice architecture" lexicon of Richard Thaler and Cass Sunstein's *Nudge* (2008), planning remains the cultural, organizational, and operational default for too many enterprise leaderships. Sustainable cultures of experiment require healthy conversations—and arguments—around hypotheses. But the planning predisposition and paradigm predominates.

Hypothesis, all too ironically, becomes more afterthought than inspiration.

Turning ordinary managers into extraordinary scientists misses the point. Facilitating collaboration around simple but compelling business hypotheses is what shifts managerial momentum. Culture change begins when people recognize that hypothesis and experimental design demand profoundly different interpersonal interactions than proposals and plans do.

WHAT PROMPTS CHANGE?

Making hypothesis a habit is hard. But getting to hypothesis becomes easier when management acknowledges what it doesn't know and needs to actively learn. Getting to hypothesis becomes practical when management recognizes that simple experiments can generate better insights faster than sophisticated analyses can. "Getting to hypothesis" becomes real when a management decides that learning from experimentation is a better investment than learning from experience.

At one giant Chinese telecommunications company, first-cut 5×5 portfolios were devoid of "let a thousand flowers bloom" intrapreneurial innovation. Although its managerial ranks were filled with scientists and engineers (many of whom successfully completed graduate work in the West), the 5×5 teams produced proposals that were virtually indistinguishable from new-product business plans. Compelling hypotheses were scarce. The company was so big and unwieldy that getting people to think in terms of simple yet scalable experiments was inherently awkward. "Big companies like big projects, not little experiments," explained a senior manager.

What changed minds here? No profound epiphanies or breakthroughs. The "aha" emerged from innovation introspection. As the various teams reviewed their more grandiose plans, a few of the more serious participants agreed that there were important things their teams didn't know about their business or their customers.

These weren't marginal gaps or missing pieces of business puzzles; these represented fundamental questions that the entire organization—not just a function or two—needed to understand better. Their grand plans had huge holes.

At those moments, rapid-fire arguments in Mandarin began. Via translators, I heard various senior managers suggest studies and analyses and committees. But others seized the opportunity to frame a business hypothesis around their admitted ignorance.

The conversational focus shifted from the new plans that they thought they wanted to implement and toward new realizations around what they needed to learn. Instead of debating, "How could we make this plan even better?" they began asking, "What do we need to learn—as soon as we can—to make this work? What kinds of experiments might give us those understandings?"

The shift from improving a plan to increasing understanding and awareness in a measurable way made the difference. For this telecom, creating better user experiences via smartphones and other digital devices became the conversational battleground for competing hypotheses. The company's innovation leadership didn't want to just imitate how Western companies defined user experience—they wanted to offer UXes that reflected Chinese values (as they saw them).

That became a hotly inspirational theme for innovative hypotheses and experimental design. Two groups proposed hypotheses to identify possible segments of urban users based on gender. Another group devised a user experience (UX) portfolio of hypotheses based on redefining personal customization. A customization road map of future features and functionality turned into a constellation of hypotheses that lent themselves to rapid testing.

A more controversial hypothesis dealt with what kind of a collaborator/innovation partner the company should be with local and regional entrepreneurs. Framing those themes as testable hypotheses rather than as executable plans profoundly altered team dynamics. Hypotheses, not programs or projects, became the organizing principles for discussion.

Via translation, I heard managers begin competing to design experiments to better understand their fundamental business challenges. As is often the case, those discussions iteratively led to revising and refining hypotheses. Those iterated hypotheses led directly to new experimentation designs.

A crucial element in 5×5 teams' willingness to explore hypotheses and experiments was the fact that, as a telecommunications giant, the company fully grasped the power of "network effects." The Internet clarified that market reality.

Technologists and marketers alike understood how simple experiments in smaller towns and cities could scale technically to serve regions. Telecommunications networks lend themselves to scalable experimentation of services and UXs. In other words, the economics of bottom-up innovation experiments were as good as—or even better than—the efficiencies of top-down strategic launches. But that was neither the culture nor the practice of the company.

"We review plans, not experiments or hypothesis," said one business unit leader. "We are not used to this."

The contrasts between grand plans and granular experiments usefully altered expectations. The simplicity, accessibility, and possibility of experimentation made the potential business value of the hypotheses feel real.

Participants started challenging themselves and each other. Did their large, lumbering firm actually have the ability to perform these business experiments relatively quickly and cheaply? Would they learn more from experiments or from surveys and analyses? Would learning from experiments be safer and more economical than learning from experience?

Needless to say, that's exactly the innovation discussion the company's top management had said they wanted. Top management wanted its teams to become nimbler and more agile at exploring innovation opportunities.

The same perceptual and behavioral transitions occurred at a Brazilian cosmetics company. Initial 5×5 brainstorming focused on identifying new ideas for customers and the sales force. That changed when participants acknowledged the firm's fundamental ignorance about what separated their very best salespeople from the pretty good ones. For example, could the company's best customers be turned into influencers, sales partners, or even into salespeople themselves? Should they? These were core cultural and strategic questions that everyone in the room cared passionately about. Those questions quickly became testable hypotheses.

"We're Brazilian. We do like to pilot and test new products with our salespeople," noted one research and development (R&D) executive at the company, "but we would never say that we are doing experiments with them or for them. ... This is a different way of working with them."

From past experience, the company knew that reliable answers wouldn't come from surveys and questionnaires. Simple experiments to test the impressive variety of hypotheses they generated, however, might offer insights they'd never imagined. The Brazilians had little difficulty framing "passionate curiosity" as business hypothesis whenever they reached rough consensus on what they wanted to discover. Unsurprisingly, team members found themselves caring about the experiments they designed. This was more than an intellectual exercise.

While their industries, cultures, and companies were profoundly different, the two firms got to hypothesis in roughly the same fashion: planning mattered less as learning mattered more. A plurality of 5×5 teams agreed that designing portfolios of simple experiments could be more practical than conjuring grand plans. Hypothesis trumped planning as a collaborative tool for surfacing actionable innovation insights. Good experiments could make good plans better. And 5×5s began to change minds by beginning to change behaviors.

SEARCHERS VERSUS PLANNERS

New York University development economist and former World Bank senior research economist William Easterly offers a clear and brilliant delineation of these rival approaches in a different but compelling global context. In his 2006 book *The White Man's Burden: Why the West's Efforts to Aid the Rest Have Done So Much Ill and So Little Good,* he divides the international aid community into two culturally distinct groups: planners and searchers

"A Planner thinks he already knows the answers," Easterly writes. "A Searcher admits he doesn't know the answers in advance; he believes that poverty is a complicated tangle of political, social, historical, institutional, and technological factors." Planners trust outside experts. Searchers emphasize homegrown solutions.

Local knowledge, local skills, and local circumstances, Easterly argues, are integral to sustainable success. In one example, he details anthropologist James Ferguson's case study of a failed Canadian International Development Agency/World Bank initiative to help farmers in the mountains of Lesotho gain market access and develop modern methods of livestock management and crop production. "The project," he notes, "promised to increase food yields by 300%."

But as social critic Virginia Postrel recounts in her review:

… it was a complete flop. The range-management techniques conflicted with local law, which guaranteed open grazing, and the farming plans were doomed by the region's bad weather.

In fact, the locals already knew the area wasn't good for farming. "The project managers complained that the local people were 'defeatist' and didn't 'think of themselves as farmers,'" Easterly reports. "Perhaps the locals didn't consider themselves farmers because they were not farmers—they were migrant workers in South African mines."

Failure is, of course, part of trial-and-error learning. The problem is that aid programs rarely get enough feedback, whether from competition or complaint. Instead, Easterly notes, advocates measure success by how much money rich countries spend. Praising the G-8 industrialized nations for doubling aid to Africa, he says, is like reviewing Hollywood films based on their budgets.

So, what is Easterly's Big Idea? "The only Big Plan is to discontinue Big Plans," he declares. "The only Big Answer is that there is no Big Answer. People everywhere, not just in the West, can all be Searchers."

What is Easterly's most important recommendation in the global fight against poverty? "Experiment. Evaluate based on feedback from the intended beneficiaries and scientific testing."

Of course, foreign aid investment in impoverished nations isn't directly analogous to business investment in innovation. But Easterly's "Planners" and "Searchers" personae seem less like caricatures than transcendent organizational stereotypes. After all, why would experienced planners bother crafting hypotheses or planning experiments? Ostensibly, they know what they're doing. "If we build it, they will come" is less a business hypothesis than the most expensive possible planning assumption.

Where plans are based on a presumption of knowledge, experimentation—searching—is based on a presumption of ignorance. The difference couldn't be starker. Yes, most organizations have a mix of planners, searchers, and everything in between. But the empirical reality remains that the resources devoted to plans and planning vastly exceed those for hypothesis and experimentation.

That's why an entrepreneur like David Kelley, the founder of IDEO, America's largest industrial design firm, can tongue-in-cheekily observe, "At IDEO, we believe that enlightened trial and error beats the planning of flawless intellects."

Of course, IDEO and its global clientele make plans and strategies. But enlightened design thinking means that they're subordinate to trial verdicts and error correction before, not after, implementation. But Kelley's comment amusingly confirms the sociocultural reasons why planners typically enjoy primacy in the enterprise. The Planning Fallacy is alive and well and dominating the innovation agenda in the executive suite.

Hypothesis is suspect precisely because it is an admission of ignorance. People experiment not because they know what the outcome will be, but because they don't. An authentic commitment to truly effective hypothesis and experimentation requires character traits that planners with flawless intellects understandably find uncomfortable. Curiosity and humility come less naturally to people who already know the answers or what the strategy must be. Contrary evidence or ambiguous results are unwelcome. Stick with the plan. Stay the course. Keep implementing. We can always adjust if we have to.

Serious experiments, by contrast, demand both the curiosity to undertake them and the humility to be open to their outcomes. Experimenters accept the possibility—even the probability—that what they learn might change their minds and direction. And, unlike comprehensive plans designed to address every possible contingency in advance, there's always another hypothesis to test and experiment to run. That relentless curiosity and constant humility needs a sense of urgency. Learning more faster becomes a professional constant.

That combination of healthy curiosity, humility, and urgency explains why hypothesis and experimentation aren't overrepresented in executive session. The majority of managers want better answers, not better questions; they want to project greater confidence, not display greater humility; they want to demonstrate superior knowledge and expertise, not reveal their ignorance; and they want to urgently implement, not experiment. Hypothesis and experimentation are perceived as creating friction and heat, not lubrication and light.

That's what really makes hypothesis culturally, organizationally, and operationally difficult. And 5×5s are only the first step toward overcoming the cognitive and cultural resistance that dominate managerial mindsets.

OVERCOMING RESISTANCE

But I've consistently observed three heuristics for hypothesis that improve the odds for executive acceptance and assistance. They've emerged from witnessing how 5×5 teams successfully navigate the transition from planners to searchers (and some never do).

The Business Hypothesis Frames a Challenge That Executives Care About

Emotional appeal and strategic alignment matter enormously. A business hypothesis that resonates with a core value, a competitive advantage, or a declared strategic imperative always commands attention. For the Chinese, exploring how simple changes in UX might reduce the cost and increase the opportunities for segmenting their customers was enormously appealing. The company could always do a study. But it recognized that it could learn even more by running simple experiments. The business case was enhanced by executive curiosity at multiple levels.

For the Brazilian cosmetics firm, turning the company's best customers into evangelists inspired a multiplicity of hypotheses. How should the "best customers" be defined and qualified? Should they be trained or educated? How should they be compensated? How could they work with the existing sales force? Everyone had exciting ideas and proposals. The challenge of collaboratively converting them to testable hypotheses suggesting quick, cheap, and simple experiments proved a powerful constraint. But everyone understood that those experimental insights would constructively inform the firm's formal planning process. Culture changes when executives care as much about their hypotheses and experiments as they do about their plans. You have to care about both.

The Business Hypothesis Provides an Organizing Principle for Convening a Team

As undeniably important as passion and executive support may be, the most effective hypotheses have a cross-functional appeal that inspires authentic collaboration. The most influential hypotheses don't just excite marketing or sales—they get the folks in human resources (HR), customer service, and R&D ihot and bothered, too. The issues and concerns hypotheses raise become reasons and requirements for multiple functions to collaborate on experimental design.

Indeed, 5×5 teams typically discuss how important their multiple perspectives are in crafting hypotheses with enterprise appeal.

For the Brazilian cosmetics company, marketing, sales, information technology (IT), customer service, and HR all wanted and needed to be involved. "We understood from the beginning that this hypothesis was too big to be controlled by Marketing or Sales alone," said the R&D executive. "It made sense to first do experiments that everyone would be interested in and involved in."

This cross-functional, more collaborative design ethos helps prevent an important hypothesis from being seen as the captive of vertical functions like marketing or R&D. Broader appeal facilitates broader organizational contribution, coordination, and support.

The Business Hypothesis Can Identify, Incorporate, and Influence Adoption

This heuristic relates directly to connectivity, scalability, and impact. The effective hypothesis doesn't invite management to set up a separate skunkworks or hive off a dedicated team. Participants can see easily how the experiments flowing from the hypothesis can have a direct impact on the business. For the Chinese telecom, the UX experiments connected directly to the company's strategic marketing, branding, and service aspirations. People understood how experimental insights would transition to the network and its planned services. For the Brazilian company, 5×5 participants recognized that the goal wasn't producing provocative experiments—it was to create experiments that would inform all the relevant functions in a constructive way about how best to proceed. Experiments weren't supposed to take place as a sideline; they were supposed to generate actionable insights for everyone involved in their design.

The measure of effectiveness of an experiment wasn't just whether it successfully tested the hypothesis. Rather, it was how well it influenced the business's ability to identify and incorporate the relevant results. Again, these heuristics highlight the central observation that culture, not cognition or managerial capability, represents the most serious obstacle to investing in hypothesis and experimentation. But the economic and organizational payoffs to building competences around crafting simple hypotheses are enormous.

14 Q & A

**YOU'VE DONE LOTS OF THESE.
WHAT ARE THE THREE MOST IMPORTANT THINGS
THAT YOU'VE LEARNED FACILITATING 5×5s?**

The top-line takeaway has to be that almost everybody finds the effort well worth their time and effort. I've rarely run across organizations that felt they could have done just as well setting up an innovation task force or hiring an external consultant.

But the three more detailed learnings that matter most are:

1. Crafting compelling hypotheses is hard.

I can't stress this enough. It generally takes more effort than teams are comfortable with to craft simple, interesting, and important hypotheses that get people excited. For whatever reason, thinking in terms of causal links—that is, "If we do X, then our customers [or suppliers or partners] will do Y" —turns out to be challenging. I get lots of questions from teams trying to sharpen and refine their hypotheses. My immediate advice is to think in terms of themes—like UX or segmentation or social media—and ask the fundamental business questions that matter most about those themes. Then convert the best of those questions into a hypothesis.

Let's say, for example, that the theme is "improving the capabilities of our customers." A logical question might be, "What apps could customers download that would dramatically improve their ability to get more value from our products?" The beginnings of a 5×5 hypothesis might look like, "If we offer our best customers apps that do X, we'll measurably increase our ability to upsell and support our highest-value features." Of course, a serious 5×5 team will identify whether marketing, sales, customer support, or information technology (IT) experts should offer the app. Similarly, which 'best' customers should get the offer, and how? What level of specificity is likely to generate the best learning fastest? Upon reflection, would we be better off providing apps to our typical customers instead of our best customers? These questions make hypothesis design hard.

2. Everyone likes looking smart in front of the boss.
Without question, the secret sauce of 5×5 effectiveness is the opportunity to present an impressive experiment portfolio before top executives (and even board members). The 5×5 both pushes people to polish their presentational skills and have bosses pay particular attention to which teams have talent. Is there an *American Idol* or *Shark Tank* or *Dragon's Den* element to these presentations? Yes. But the overwhelming majority of 5×5 participants want their colleagues, bosses, and peers to say, "You know, that's a really important experiment. I would have never thought about it that way. In fact, your entire portfolio changes how I look at things. Can we talk later?"

3. Truly collaborative and diverse 5×5 teams do best.
Yes, there are always truly gifted individuals with truly brilliant hypotheses and even better experimental designs. They're always easy to spot. They're the ones whose 5×5 teammates say little and

do less. Their work is typically excellent. But it is firmly rooted in their individual brilliance. For some firms, that brilliance is more than good enough.

That said, the most successful 5×5 teams I've seen are genuinely diverse—they represent a cross section of the enterprise and have a mix of experiences. Everyone makes everyone on their team smarter; they complement each other.

Invariably, their portfolios and experiments reflect their diversity—they're more holistic, they address issues the entire organization cares about; the experimental designs cut across functional boundaries. Top executives come away with a sense of how their organization really operates instead of what the org chart says. It's fun to watch.

WHAT'S THE MOST COMMON MISTAKE YOU SEE 5×5 TEAMS MAKE?

Immediately trying to come up with five great experiments.

The team spends most of its initial time asking each other for their best ideas rather than discussing what matters most to the organization's future. People will still have their best ideas. But it's important for a team to create some sort of shared sensibility about what they collectively want to accomplish. Successful teams commit to that; less successful teams find that shared sensibility elusive. The beauty of well-run 5×5 teams is that they collaboratively build upon and improve even the best individual proposals. It's a cliché, but there's a profound difference between a group of five people and a team. Don't dismiss or minimize that aspect of the 5×5 experience. The social side of experimentation, innovation, and presentation is important.

THIS ALL SEEMS REASONABLE AND LOGICAL. SO WHAT'S THE MAIN REASON THAT AN ORGANIZATION REJECTS THE 5×5?

Two reasons account for three-quarters of the polite rejections and refusals to go forward. The first is that the organization—or more accurately, a particular slice of leadership —has an innovation process in place, and the 5×5 is seen as either a rival or a distraction. Ironically, the 5×5's great strength is its greatest weakness here: because it's designed to be very fast, very cheap, and very simple, success might make other innovation initiatives look slower, more expensive, and more complicated by comparison. Of course, the 5×5 should complement, not replace, what's there. But why take the risk?

The other reason is simpler albeit slightly cynical: people understand exactly what the 5×5 represents. They have zero desire to confront the organizational pathologies and dysfunctions a 5×5 would surface. They know they have issues but fear a 5×5 will merely highlight them instead of being an important step to improving them. I will self-servingly note, however, that the 5×5 tends to be most welcome in organizations that have vibrant and healthy innovation cultures. Vibrant and healthy innovation cultures are almost always up for trying something new if the value proposition is there. When healthy innovators pass on a 5×5 opportunity, it's generally because they've got a full plate.

SO IS THERE SOME TRICK OR GIMMICK THAT YOU CAN SHARE FOR A 5×5 TEAM THAT WANTS TO LOOK GOOD?

Yes. At risk of sounding cynical, ignore customers. Ignore clients. Ignore partners. Identify your boss's boss or the top executive everyone knows is set for promotion next year or, of course, the CEO, and come up with a portfolio with five experiments that she'll love because they'll make her look good. You never go wrong if you can make your real customers happy.

EXPLORING
THE 5×5
EXPONENTIAL
FUTURE

15 EXPERIMENTING WITH EXPERIMENTATION

Machines are for answers; humans are for questions.

—Kevin Kelly

A first-rate laboratory is one in which mediocre scientists can produce outstanding work.

—Patrick M. S. Blackett

By design, 5×5 frameworks embrace flexibility, adaptability, and off-the-shelf technology: whatever works. But does that simple ethos hold up as technical disruptions intensify and business challenges increase? Customers crave getting more for less. Seemingly strategic competitive advantages appear fleeting. Can 5×5 initiatives successfully drive lightweight, high-impact value creation in environments rife with tomorrow's risks?

Yes. The innovation future will prove 5×5 friendly. Experimentation's emerging economics seductively favor innovators prepared to move fast. Vision remains vital. But strategic advantage increasingly relies upon diversified portfolios of innovative hypotheses. However, the customer's vision and strategic advantage matter most of all. Smart customers will see your experiments as potential investments into their own efforts to manage innovation and risk. Leaner, faster, but simpler experimentation will reinvent tomorrow's innovation ecosystems.

Tomorrow's trends are the 5×5's friends. The experiments-driven future—and future-driven experiments—will look like this:

Instant messaging. Instant search. Instant surveys. Instant collaboration. The future of business experimentation is instant. Running an experiment will be as natural and casual as performing a search. Innovators and entrepreneurs will experiment with the same frequency, immediacy, and impulsiveness that they now text and tweet. With but a few keystrokes, inchoate hypotheses will become tightly targeted A/B tests worldwide. The quicker the preliminary results, the faster they're privately tweeted out to colleagues, collaborators, and clients.

Facebook. LinkedIn. Yammer. Jive. Twitter. Pinterest. Yelp. The future of business experimentation is social. The power to self-organize and peer-review experimental designs simultaneously ensure awareness and alignment around insights. Collaborative experiments seamlessly become multimedia and multimodal. Shared visualization shapes interpersonal interaction as much as sophisticated quantification. Social networking cross-functional experimentation facilitates cross-functional experimentation in social networking.

Amazon's books. Spotify's music. Netflix's movies. The future of business experimentation is recommended. Recommendation engine technologies suggesting what books to read, songs to hear, and films to see will also suggest what business hypotheses to test. "Marketers like you," salespeople like you," or "software developers like you" will get data-driven recommended hypotheses based on statistical correlates and patter matches. Aspiring innovators will put the most desirable recommendations into their innovation "shopping carts." No doubt, some will be given to colleagues and potential collaborators as experimental gifts.

Recommenders mathematically map the specific features and attributes of people to things—products, people, content—that might be of interest. So, where Facebook suggests possible friends based on distinctive data points (people and their data) connected by carefully calculated social graphs, Amazon recommendations benchmark baskets of products that have been purchased together in the past. Not only do both approaches deliver excellent results, they demonstrably get better the more that people use them. They deserve to be emulated.

The algorithms powering these engines—regression, naive Bayes, support vector machines, decision trees, etc.—are hardly new or novel. But their computational ability to mix and match data sets makes them digital reservoirs for generating new and novel hypotheses. The meta-tagging and behavioral analytics that make Netflix video recommendations so compelling can, in probabilistic principle and practice, be repurposed for experimentation. Instead of providing personalized recommendations for binge viewing, repurposed recommendation engines can facilitate binge hypothesizing. Which recommendations will tomorrow's 5×5 teams buy?

BACON. AM. Deep Blue. Adam. Watson. The future of business experimentation is automated. But these automated algorithms eschew the human behavior/collective intelligence components that make recommenders run. Autonomous by design, they're trained to recognize patterns and possibilities that even the sharpest humans can't—or don't—see until it's too late. They're pure machine learning and intelligence. When IBM's Watson supercomputer defeated *Jeopardy!* super-champion Ken Jennings in February 2011, the most successful human player in the show's history half-jokingly scribbled on his monitor with his "Final Jeopardy" answer, "I, for one, welcome our new computer overlords."

Jennings's self-deprecatory humor was a far cry from world chess champion's Garry Kasparov's bitterness after his humiliating 1997 loss to IBM's Big Blue. "I was not in a fighting mood," he acknowledged. "I'm ashamed at what I did at the end of this match. I have to apologize for today's performance. I'm human." Indeed.

But emphasizing that sophisticated software beat the world's best humans obscures the larger point. The biggest losers in such competitions are above-average performers who typically deliver above-average insights, analyses, hypotheses, and results. Unless they add value to those superior algorithmic intellects, they're likely worth less (or even might be worthless) as innovation partners.

Computational scientific discovery—not unlike chess—has a long, intimate history with artificial intelligence and machine learning research. In the 1970s, AM—for "Automated Mathematician"—pursued the autonomous development of mathematical theorems. In the 1980s, Carnegie-Mellon's Herbert Simon, an artificial intelligence (AI) pioneer who had won the 1978 Nobel Prize in Economics) co-developed BACON, a software program designed to rediscover scientific laws and principles. As with chess, contemporaneous technological limitations constrained success.

As with chess and *Jeopardy!,* the Watson-ification and Deep Blue-ing of world-class hypothesis and experimental design in the sciences is accelerating. The robots aren't coming; they're already here. In their 2013 article, "Towards Robot Scientists for Autonomous Scientific Discovery" (http://www.aejournal.net/content/2/1/1), a British research team asserts:

> Robot Scientists are the next logical step in laboratory automation. They can automate all aspects of the scientific discovery process: they generate hypotheses from a computer model of the domain, design experiments to test these hypotheses, run the physical experiments using robotic systems, and then analyze and interpret the results. ... We look forward to a time when Robot Scientists will commonly work with human scientists to progress the path of science.

Not incidentally, the team's prototype robot—Adam—has devised hypotheses and designed and run academically publishable experiments on yeast genetics. Adam's discoveries, according to its creators, have been modest, but not trivial. As the underlying technologies exponentially improve, Adam's technical descendants will enjoy less modest successes in many disciplines.

Talented humans don't inherently matter less in the coming experimental future, but empowered machinery inherently matters much more. Humanity has lost its monopoly on discovery and design. That means innovative organizations need to think twice about the economics of thinking twice. Recommending better automation may prove to be a superior investment over automating better recommendations. Quasi-autonomous "Adams" will participate as partners to tomorrow's 5×5 teams. Not using robots will become unnatural.

iPhones. Google Glass. Tablets. Nest. Internet of Things. Tesla. Quantified Self. The future of business experimentation is pervasive. Everyone—and everything—are becoming more interconnected or interoperable in some meaningful way. Evolving global grids of digital data and devices are only precursors to new innovation options. The rise of ubiquitous interconnectivity and interoperability creates the biggest research laboratory in the history—and future—of the world. Innovative opportunities for combinatorial experimentation with people, devices, and things will be everywhere.

Pervasive instrumentation means there's always data for collection, correlation, and contemplation. Everything is subject to creative hypothesis and experimentation. Metaphorically speaking, smart phones are sensors and tablets are test tubes. The right apps can turn digital devices into microscopes, telescopes, oscilloscopes, and spectrometers instantly. Social media can sample sentiment locally, regionally and worldwide; search engines can diagnose possible pandemics of people, plants, or livestock.

Mashing up ever-smarter and self-aware technologies creates the need for new measures and metrics to assess what's really going on. Passive observation doesn't go far enough; innovators will explore actively how novel interconnections can facilitate new value creation.

"Like many technologies," says Andy Hobsbawm, a founder and chief marketing officer of EVRYTHNG, a British "Web of Things" software company, "the internet of things is best considered from the perspective of adoption rather than purely invention. How will people interact with the new innovations they are supposed to benefit from? If passively—by having useful services invisibly accomplished for them—then what input should they have in how this happens? If actively engaging in how these services are carried out, then what physical and digital interfaces do they need?"

The answers to Hobsbawm's questions cry out for innovative hypotheses and experimentation. But these Internets and webs of things—be they consumer or industrial—are intrinsically and inherently platforms for rapid experimentation and testing. Interconnectivity and interoperability make them de facto laboratories, replete with state-of-the-art instrumentation. Experimenting is the enzymatic middleware investment that turns those laboratories into innovation engines. Pervasiveness changes the game.

Instant. Social. Recommended. Automated. Pervasive. These are the ingredients, raw materials, and components that will shape the future cultures and capabilities of innovative business experimentation. But don't think of these words as individual elements of experimental design. Think of them as expressing organizing principles for fundamental value investment in innovation.

How will—or should—organizations invest in "instant" and "social" to differentiate the value that they can get from experiments? What recommendation genres provide more reliable experimental insights than automated systems? Which investment criteria do the organization and its innovators want to own as they experiment? The future impact of 5×5s, inside the enterprise and out, depends on these investment choices.

These five secular trends reflect a fundamental shift in how organizations worldwide manage their innovation organization and investment. This architectural shift influences virtually every sector of the global economy. That technical architecture has transformed the economics of innovation and will do so for decades. There's no escaping its pervasiveness and potential. Innovate. Iterate. Repeat.

For well over a century, the dominant innovation paradigm was defined by research and development—R&D. Linear, capital-intensive, and complex R&D processes—carried out by companies ranging from DuPont to Bell Labs to Xerox to Sony to General Motors (GM) to HP to General Electric (GE) to IBM to Procter &

Gamble (P&G) to Siemens—organizationally defined big-company innovation. The biggest companies typically invested the most in R&D.

Short of acquiring innovative companies or technologies, R&D was the lens and prism through which innovation investment was viewed. The costs and challenges of scaling basic research into practical development and pilots were extensive, expensive, and fraught with risk. The economics of innovation were unkind. Commercializing innovation busted both budgets and schedules.

That traditional, linear R&D paradigm is dying. The increasingly networked economics of experimentation is killing it.

This rise of networked organizations—the "networkification"* of enterprises, both global and local—creates novel, nonlinear, and iterative innovation options. The enterprise innovation paradigm is shifting from R&D to E&S: that is, from research and development to experiment and scale. The ability of networks—especially digital networks—to quickly and seamlessly scale small experiments into new services represents as radical an innovation revolution as Henry Ford's Highland Park assembly line.

Where factories exploit economies of scale through operational efficiencies and greater production volumes, innovators exploit network economics by scaling simple experiments into value-added services. The most successful representatives of this innovation sensibility are, of course, the Amazons, Intuits, Googles, Facebooks, Airbnbs, Netflixes, and Ubers. They are leading experimenters in network effects. Their ability to both scale experiments and experiment with scale is core to their operational effectiveness and user experience. One experiment leads to another. They iterate their experiments and experiment with their iterations.

---

* As the Internet bubble inflated in 1995, I half-jokingly suggested a "Schrage's Law of Networks" to complement the more familiar Moore's Law and Metcalfe's Law: "The surest way to add value to a network is to connect it to another network." This half-joke turns out to be largely true.

But the broader and more compelling economic insight is that the more net-centric the enterprise, the more cost-effective innovation investments that are net-centric become. For example, the more financial service companies—like banks and mutual funds—rely upon and resemble networks, the easier and cheaper that experiments with customer recommendation engines or credit card loyalty points become. The more creatively that retailers explore omni-channels to link physical and digital shopping experiences, the more opportunities materialize for cheap experiments with "path-to-purchase" shoppers. As logistics providers embrace multimedia networks to track multiple supply chains worldwide, experimenting with innovative allocation, aggregation, and routing algorithms becomes faster, cheaper, and simpler.

In global industry after global industry—health care, construction, education, and professional services—investments in network interconnectivity and interoperability inherently make targeted investments in hypothesis and experimentation more valuable. Networkification shifts experimentation from a variable cost to a marginal one. Networkability literally creates new markets for experimentation. As businesses around the world accelerate into the cloud, the global innovation climate changes.

"The infrastructure is coming into place to minimize the cost per test and maximize the number of trials per year," asserts Jim Manzi, the cofounder and chairman of Applied Predictive Technologies (APT), one of the most successful firms helping companies design and run field experiments. "Our abilities to exploit segmentation to better generalize the results of experiments are also increasing."

Cloud computing enables what might be called "accordion innovation": a 5×5 team can try something small, test it at a larger scale, and then zoom it back down to refine what didn't work in an iterative way. Scale can be customized as easily as the experiment. That means cloud computing's "software-as-a-service" model effectively offers experimentation as a service. That's huge. That will command C-suite attention. "We're going to see greater executive commitment to apply and act on experiments for the questions that have the greatest potential for shareholder value," Manzi predicts. These experimentation economics recalibrate corporate expectations around tactical and strategic financial investment.

Q & A: THE EXPERIMENTER

Gary Loveman, an economist who became CEO of Caesars Entertainment in 2003, demands that his employees operate the business by analyzing data rather than leaning on hunches. Below are excerpts from a Q&A he and I did in 2011 for *MIT Technology Review*.

WHAT'S THE MOST IMPORTANT THING ABOUT CAESARS' CULTURE OF EXPERIMENTATION?

We need to overcome hunch and intuition with empirical evidence. We've set up a process and a discipline for evaluating our intuitions and improving our understanding of what our customers prefer. We can start with a hunch or strong belief, but we act on it through experiment. ... We've gone from the introduction of experimentation as a technique to a culture of experimentation as a business discipline. ....

WHAT'S BEEN THE BIGGEST SOURCE OF RESISTANCE?

Impatience and risk aversion. Let's say that one of our properties had lower revenues than they'd like, and they think they know the reason why. Instead of running an experiment to test that reason, they don't use a control group and pollute the entire process. This impatience and hubris breaks the discipline I want us to have. A well-designed experiment is the better way of testing that reason and learning what matters.

WHAT MAKES SO MANY EXECUTIVES PREFER TO RELY
ON THEIR EXPERIENCE AND ANALYSIS OVER SIMPLE
EXPERIMENTS?

There's a romantic appreciation for instinct and, frankly, an absence of rigor for the application of more scientific approaches. What I found in our industry was that the institutionalization of instinct was a source of many of its problems.

WHEN YOU GOT YOUR ECONOMICS PHD FROM MIT
IN 1989, BEHAVIORAL ECONOMICS AND EXPERIMENTAL
ECONOMICS HAD A MIXED REPUTATION. THEY NOW
SEEM TO INSPIRE HOW MANY BUSINESSES AND INDUS-
TRIES TRY TO INNOVATE?

My impression is that when I got my PhD, we were really manipulating mathematics for our own amusement, and we weren't producing all that much to help real people make real decisions. That was dissatisfying to me and, frankly, frustrating. ... Of course, with *Freakonomics* and *Predictably Irrational* these themes have become more popularized and accessible. It's a very heartening development, and it's increased my enthusiasm for my own discipline enormously.

WHAT DO YOU LIKE TO TELL YOUR ACADEMIC
COLLEAGUES ABOUT THE CHALLENGES OF REAL-WORLD
EXPERIMENTATION AND INNOVATION?

Honestly, my only surprise is that it is easier than I would have thought. I remember back in school how difficult it was to find rich data sets to work on. In our world, where we measure virtually everything we do, what has struck me is how easy it is to do this. I'm a little surprised more people don't do this.

Cloud facilitates bottom-up experimentation and innovation just as readily as it empowers top-down innovation initiatives. Pick how small you want to start, how big—and fast—you want to scale, how much you need to learn at each level, and how many iterations that requires. E&S increases innovation velocity. Instant. Social. Recommended. Automated. Pervasive.

But networking networks offers the greatest potential for breakthrough hypotheses and experimentation that drive disruptive innovation. Vint Cerf, the godfather of the Transmission Control Protocol/Internet Protocol (TCP/IP) that makes the Internet a network of networks, recalls, "We thought we were building a system to connect computers together. But we quickly learned that it's a system for connecting people." In reality, it's both. Interoperability becomes, as I observed elsewhere, innovation's great enabler:

"More innovators in more disciplines are investing more in interoperability as both a business and research strategy. Nascent nanotechnologies are being mashed up with biotechnologies. Facebook pages mash up with Global Positioning System mobile phones. Rechargeable batteries can mash up with programmable solar cells. Seemingly disparate devices and disciplines that ordinarily would have zero interest in interoperating creatively, or zero capacity to do so, might find novel relationships cheap and easy. Successful interoperability dramatically cuts the costs, risks, and complexities of hooking up. Barriers to interdisciplinary innovation tumble.

"How might interoperability between Siemens cochlear implants, iPhones, Nike running shoe accelerometers, LG microwave ovens, Xboxes, and BMW sedans create bold entrepreneurial, or diversification, opportunities? Who knows? But ... that the question piques curiosity reveals fundamental changes in the global innovation climate."

Multidisciplinary and interdisciplinary opportunities for combinatorial experimentation and innovation continue their exponentially expansion. The most important takeaway: Favorable economics of interoperable innovation will tempt ambitious "inter-preneurs" to test their ideas.

Nimble and cost-effective capitalization of interpreneurial inno-vation is unlikely to come from traditional R&D models. Network economics render them cumbersome anachronisms. Innovation successes will emerge from the best of decentralized and distrib-uted E&S initiatives—portfolios of experiments that smoothly and smartly scale across internets of people and things alike. Instead of 5×5 experiments, tomorrow may well bring 5×500 or 5×500,000 experiments for running and review.

But who exactly will design and deploy E&S? What tools and technologies will support 5×5 interpreneurial efforts? While the Deep Blues and Watsons of E&S are well on their way, George Mason economist Tyler Cowen suggests the centaurs of "freestyle chess" as innovation inspiration. Freestyle chess tournaments give humans unrestricted use of computers during matches. These man/machine teams are commonly called *centaurs*. In his book *Average Is Over,* Cowen argues that the future economic value of knowl-edge workers is intimately dependent upon the future capabilities of machine intelligence and learning. Chess is both his metaphor and model. (Not incidentally, Cowen was once, at 15, the New Jersey state's youngest chess champion.)

He points to research suggesting that "a striking percentage of the best or most accurate chess games of all time have been played by man-machine pairs." Yes, programs like Deep Blue can beat the best grandmasters. But smart humans aided by smart chess soft-ware can frequently beat machines alone.

The catch? Truly effective man/machine collaboration requires modesty as much as intellect. "The really good human players are too tempted to override the computer and substitute in their own judgment," Cowen observes. "The best freestyle teams ... are quite epistemically modest. ... And what they are really good at is asking questions. So they'll run two or three different computer programs and then just check on where do those programs disagree. And then they'll probe more on those points. That's what the humans do well that the computers, at least not yet, aren't able to copy. So it's knowing what questions to ask that has become the important human skill in this freestyle endeavor."

Epistemic humility, says Cowen, must become a knowledge worker core competence as machine intelligence comes to rival their own. "Think in terms of this future middle-class job," he adds. "You read medical scans, and you work alongside a computer. The computer does most of the judging, but there are some special or unusual scans where you say, 'Hmm, that's not quite right—I need a doctor to look at this again and study it more carefully.' You'll need to know something about medicine, but it won't be the same as being a doctor. You'll need to know something about how these programs work, but it won't be the same as being a programmer. You'll need to be really good at judging, and being dispassionate, and you'll have to have a sense of what computers can and cannot do. It's about working with the machine: knowing when to hold back, when to intervene."

Far more than freestyle chess tournaments, this workplace scenario maps perfectly to the challenge of crafting novel hypotheses and designing cheap, ingenious, yet simple business experiments. Data will be digital, patterns and positions complex and competing. But humans—5×5 teams—need to do "freestyle experimenting" as competently and creatively as winning centaurs play chess.

Cowen's humbling conclusion? "So, wisdom and modesty will become much greater epistemic virtues in the future scheme. I think that's overall a good thing. We *should* revere those qualities more. And we will have to, looking forward."

Making virtue of necessity is what dynamic markets do. But as Cowen's centaurship admonitions suggest, profitably transforming necessities into virtues both challenge and change cultural expectations. This book needs to end as it began: explicitly acknowledging that experimentation is as much a cultural choice as a technical competence. As novel experimental techniques and technologies co-evolve, so do the people and cultures employing them.

The essential concern is that what organizations now understand as the scientific method is changing, and so is the culture of scientific inquiry and insight. This recalls the physicist Max Planck's cynical cultural insight into scientists' search for truth. "A scientific truth does not triumph by convincing its opponents and making them see the light," he wrote, "but rather because its opponents eventually die and a new generation grows up that is familiar with it."

Perhaps that's too extreme. But the secular trends shaping tomorrow's business experimentation are shaping scientific experimentation as well. Might business and science be on convergent paths? Who knows? But their similarities are rapidly becoming as important as their differences. Kevin Kelly, a founding editor of *Wired* magazine and author of the brilliant *What Technology Wants*, describes how science's ongoing evolution redefines not just the nature of experimentation, but the culture of its experimenters. Provocative nature/nurture debates are breaking out around the future of scientific insight.

"[L]ots of things that we assume or we now associate with the scientific method were only invented recently," Kelly observes. "… some of them only as recently as 50 years ago—things like a double-blind experiment or the invention of the placebo or random sampling were all incredibly recent additions to the scientific method. In 50 years … the scientific method will have changed more than it has in the past 400 years, just as everything else has.

"So the scientific method is still changing over time. It's an invention that we're still evolving and refining. It's a technology. It's a process technology, but it's probably the most important process and technology that we have, but that is still undergoing evolution refinement and advancement and we are adding new things to this invention. We're adding things like a triple-blind experiment or multiple authors or quantified self where you have experiment of N equals one. We're doing things like saving [and transmitting] negative results. There's many, many things happening with the scientific method itself—as a technology—that we're also improving and … will affect all the other technologies that we make."

But the knock-on and multiplier effects that Kelly identifies about the future of scientific methodology have a striking common narrative about their human interlocutors. "Every time we use science to try to answer a question, to give us some insight, invariably that insight or answer provokes two or three other new questions," he says. "Anybody who works in science knows that they're constantly finding out new things that they don't know. It increases their ignorance ... while science is certainly increasing knowledge, it's actually increasing our ignorance even faster. So you could say that the chief effect of science is the expansion of ignorance."

Substitute *business* for *science* and *opportunity* for *ignorance,* and Kelly's tongue-in-check observation explains why the ongoing revolution in experimentation defines the innovation future. There's never been a better time for innovating with experiments or experimenting with innovation.

APPENDIX: THE INNOVATOR'S HYPOTHESIS MATH

WITH PROFESSOR EVA K. LEE
Director, Center for Operations Research in Medicine and HealthCare
H. Milton Stewart School of Industrial and Systems Engineering
Georgia Institute of Technology

Be bold. Be creative. Be frugal. But, whatever else you do, be sure to play the odds. Make probability your friend. This appendix discusses the simple math that makes 5×5 portfolios a smart and superior innovation investment.

A brief appreciation of decision theory is central to understanding why experiment portfolios work. This discipline enjoys a rich tradition of real-world mathematical success spanning from Pierre Fermat and Blaise Pascal to Nobel Prize–winning economics. Acquiring mastery is difficult and challenging; even decision theory experts disagree about the best way to use their work. But the field's complex history and technical sophistication shouldn't obscure its more accessible and applicable insights. This section makes a mathematical case for why handfuls of targeted experiments frequently prove much more valuable to innovation investors than carloads of traditional quantitative analyses. Call it "Decision Theory Made Simple"—or, more accurately—usefully simplistic.

The most important technical concept to grasp is the connection between expected value and subjective expected utility (SEU). That is—given expected or known risks—what are the most reasonable and rational outcomes that a decision maker will get from making a choice or investment? More simply, if a decision is a bet, what's the expected payoff?

Yes, this is decision theory as formalized gambling. The intellectual and historical origins of this are firmly rooted in the real-world challenges of risk-taking and gambling. For example, Fermat and Pascal's fruitful collaborations in the mid-seventeenth century in the realm of dice games effectively launched probabilistic analysis. Indeed, the relationship between great mathematicians and gambling is long and intimate. The entire mathematical, psychological, statistical, and probabilistic evolution of decision theory is a story of how people choose to manage the calculated relationships between perceived risks, rewards, and regrets. In other words, what kind of gamblers, bettors, and investors do people want to be?

The simplest gambling/betting/investment equation is
Expected Value = Payoff × Probability

For example, if you bet a dollar to win a dollar on the toss of a fair coin, your expected-value payoff is 50 cents. Why? Because there's a 50/50 chance you'll bet right. If you bet a dollar to win $5, then the expected value is $2.50. Decision bets become more interesting and complicated when you're rolling dice, drawing cards, investing in stocks, or purchasing financial instruments where market conditions may mean you can lose more than you put in. Calculating the odds, the risks, and potential payoffs becomes increasingly more complex and subjective. The easily computed expected value evolves into a more complicated variable: SEU—that is, the perceived probabilities and risks become more subjective while the value or worth is expressed in terms of utility—a general measure of desirability. The "sure thing" become as mythical as unicorns and griffins.

Most decision makers want to get the maximum possible payoff for the minimum level of acceptable risk. They expect to make trade-offs between potential payoffs and possible risks. This is the decision theory counterpart to "buy low, sell high"—where different decision makers may disagree about how high is high and how low is low. That's where subjective assessments and conflicting utilities come in. After all, expected utility is a mathematical abstraction and simplification of reality.

But if we think of decision makers as investors in innovation who are making bets on the possible value/utility of experiments to generate useful information and insight, then SEU provides a powerful and useful way to tilt the odds in their favor. Well-run experiments can build confidence both organizationally and probabilistically.

The underlying computations are no more complicated than multiplication, addition, and simple division. The real challenge comes less from the application of sophisticated statistical principles than a willingness to rethink how organizations can use experiments to gain insight faster, better, and cheaper.

Consider the following thought experiment, which is firmly based in real-world innovation experience. While oversimplified and simplistic, this example highlights how simple experiments can—quite literally—add up to something that serious innovators should organizationally, culturally, and operationally explore.

A company—let's call it General Innovation—is exploring the risks, costs, and benefits of a Brave New Innovation Initiative (BNII). The organization's innovation committee leadership takes a good, hard look at the innovation brief and says, "We've seen this kind of Brave New Initiative before. Give us 100 days to really analyze the idea and its assumptions. We'll talk to experts inside the company and out; whip up the right scenarios; and run the best numbers we can. We typically need a $100,000 budget to get the job done right in the allotted time. But at the end of those 100 days, we'll give top management a thorough analysis of what's most important and how we should proceed. Because it's a Brave New Innovation, there are no guarantees. But—based on our experiences—I'm 90 percent confident that our report will give management 90 percent of the information it needs to make an informed go/no-go decision and innovation investment."

Let's recall the simple expected value equation; it effectively captures this SEU, promised by General Innovation's innovation committee:

A 90 percent confidence level for ensuring 90 percent of the essential analysis calculates to a SEU of 81 percent. (That is, 0.9 × 0.9 = 0.81.)

So General Innovation is spending $100,000 over 100 days to get roughly 80 percent of the critical information that it needs for its BNII.

But wait! There's a rival group of innovation intrapreneurs at the firm who want to take a radically different approach to the BNII. They think that they can get essential insight and information to manage the innovation risk significantly faster, better, and cheaper. While excited about the BNII's potential opportunities, they're less interested in and supportive of traditional analysis. For them, fast and cheap experimentation based on smart and savvy hypotheses offers a better way to go. They want to craft a portfolio of experiments that will deliver the essential "What should we do next?" information faster, better, and cheaper than the 100-day analysts.

They're committed to creating clever experimental designs that will let them learn a lot quickly and less expensively than the analysts. They do the 5×5 exercise, taking care to select hypotheses and experiments that deliver—in their collective professional opinion—the maximum possible insight in the minimum amount of time. No experiment can cost more than $5,000 or take longer than 20 days to run. An intriguing mix of hypotheses and experiments result:

The first experiment is the long shot: The team estimates that there's a 10 percent chance that this bold experiment will generate 75 percent of the vital information necessary to assess the prospects of the BNII with confidence. That means its SEU is 0.10 × 0.75 = 0.075.

The second portfolio experiment has better odds, but it wouldn't yield as much innovation insight. The team believes that there's a

20 percent chance the experiment will address 66 percent of the desired information. 0.20 × 0.66 makes the SEU 0.132.

Experiment 3 climbs a little higher up the probability ladder. The team estimates that it offers a 30 percent chance to capture half the required BNII insight. That makes the SEU 0.30 × 0.50 = 0.150.

The fourth portfolio experiment creates a perceived 40 percent opportunity to learn 40 percent of what the organization needs to know. Its SEU is 0.160.

The fifth and final portfolio experiment has the best odds: the experiment has a 50 percent chance of discovering 33 percent of what will make the BNII work. The SEU payoff: 0.165.

Yes, this portfolio represents a thought experiment about the potential payoffs of real experiments. For the sake of logic and consistency, I've declared that the better the odds, the lower the expected learning amount will be. After all, if there were a fast and cheap experiment with an 80 percent chance of yielding 80 percent of the desired information, that would certainly be the best, most rational, and most cost-effective investment. But, remember that these fast and cheap experiments aren't designed to be perfect: they're intended to generate useful and usable information with speed and insight.

Note that not a single portfolio experiment here represents a sure thing. Not a single experiment offers odds better than a coin toss. Individually, they don't necessarily seem like good innovation investments. Collectively, though, look what happens when you add up the SEU payoffs:

```
    0.075
+   0.132
+   0.150
+   0.160
+   0.165
───────────
```

Total: 0.682

This experiment portfolio, therefore, buys the 5×5 team an SEU of 0.682.

Now let's do a little division to go along with the multiplication and addition:

0.682/0.81 = 0.841

In other words—or, more accurately, numbers—this hypothetical 5×5 experiment portfolio buys roughly 84 percent of the promised SEU from the 100-day analysis. The 5×5 portfolio team gets over 80 percent of the innovation insight in a quarter of the time, and with a quarter of the budget.

Yes, I know that the independence of these various experiments hasn't been defined rigorously. Similiarly, this thought experiment doesn't fully account for the time and costs associated with reviewing and integrating the experimental outcomes. But the vital lessons here shouldn't be obscured by the artificiality of the exercise.

Both in theory and practice, portfolios of lightweight, high-impact experiments can generate roughly 80 percent of the desired information in roughly a quarter of the time and with a quarter of the cost of more traditional analytic and strategic reviews. Framing innovation challenges in terms of portfolios of hypotheses to be tested rather than analyses to be performed makes economic, financial, and organizational sense.

Thinking—and designing and investing—in terms of experiment portfolios creates cost-effective opportunities for iterative innovation learning and development. Why not do another round or portfolio of experiments based on what's been learned? How should we further hypothesize and experiment with the results of our hypotheses and experiments? Iterative experimentation may create faster, better, and cheaper innovation learning opportunities than do more refined analytics.

While it's undeniably true that the thought experiment portfolio presented here had an unusual array of SEU payoffs, other BNIIs might inspire hypotheses and portfolios filled with different odds and utility functions. For example, smart 5×5 teams could craft cheap experiments that had a 60 percent chance of generating 50 percent of the desired information, or an 80 percent chance of capturing a third of what's needed to know.

The vital takeaway is that experimentation is probablistically undervalued and underappreciated as an innovation insight investment by too many organizations. Bringing a diversified, portfolio theory approach to innovation insight and risk management is a demonstrably smart business investment. As one reviewer put it, "Experiments diversify and hence, offer a greater chance of knowledge. The 100-day investment is single view, so no matter how good it is in the first round, it is not sustainable in the long run, as it is too homogeneous."

But suppose that you and your colleagues can't come up with hypotheses and experiments that deliver the SEUs that you need or want? The answer is shockingly simple: Don't do the experiments!

This leads to the serious and sobering economic conclusion: There is no free lunch as far as innovation or experimentation are concerned. The experiment portfolio fails if 5×5 teams aren't collaboratively creative, ingenious, or thoughtful. The numbers don't add up if 5×5 teams do a poor job of valuing the informational insight potential of their hypotheses and experiments. Do they optimistically overestimate the utilities? Do they consistently underestimate the risks? Have they really thought through the odds of what they are trying to discover, learn, and test?

The numbers don't lie. But they do require a commitment from innovators to make them work. Experimenters can be just as overconfident and overoptimistic as the planners, of course. But, economically speaking, identifying and adjusting for that experimental overconfidence and optimism can be done faster and—in all likelihood—cheaper.

BIBLIOGRAPHIC ESSAY

This brief bibliographic essay discusses key influences on *The Innovator's Hypothesis*, as well as further suggested reading. The book draws from a variety of remarkable practitioners, researchers, and writers on innovation, risk, and experimentation. Their most important common denominator is an authentic commitment to learning by doing. These are not dispassionate creatures of analysis. The most significant differences between them, by contrast, are their attitude toward simple, fast, and frugal experimentation. There's little consensus around what is simple or fast or frugal. Those definitional disagreements and gaps create productive opportunities to experiment with experimenting.

This acknowledgment might surprise him, but Dan Ariely—the gifted behavioral economist who authored the best-selling *Predictably Irrational: The Hidden Forces That Shape Our Decisions* (New York: Harper, 2008) and *The Upside of Irrationality: The Unexpected Benefits of Defying Logic at Work and at Home* (New York, Harper, 2010)—exerted a disproportionate impact on both the 5×5 methodology and this book. When he was at MIT, I was deeply impressed with the rigor, low cost, and design simplicity that his people brought to their experiments. They relentlessly— often elegantly—zeroed in on essential questions of human behavior and choice. When reviewing the "bang-for-the-buck" benefits of behavioral economics experiments versus big-budget branding, marketing, advertising, and customer research tests run by large companies, it was no contest. The behavioral economists consistently delivered better insights faster. You may think that you're reading wonderfully informative stories about human nature, but you're actually getting a master class in experimental design. Douglas Hough's *Irrationality in Health Care* (Stanford, CA: Stanford University Press, 2014) presents a magnificent example of how bringing behavioral economic insights and heuristics to health care could—and should—inspire simple, safe, and scalable experiments that may well save both lives and money.

Jim Manzi's *Uncontrolled: The Surprising Payoff of Trial-and-Error for Business, Politics, and Society* (New York: Basic, 2012) is an epistemological tour de force. Serious, thoughtful, accessible, and provocative, there's no finer English-language survey or review of the history and culture of experimentation. No book explores and details the experimental method's strengths, weaknesses, and future potential as both a business resource and a policy tool. His mildly technical discussions about the design of experiments and randomized field trials should appeal to even the most innumerate managers, while the "The Experimental Revolution in Business" chapter affirms there's serious money to be made. In fact, Manzi is a successful experimental entrepreneur whose company, Applied Predictive Technologies (APT), is a market leader in software facilitating large-scale randomized trials. (Full disclosure: at Manzi's invitation, I spoke at APT's annual customer event and consider him both a friendly acquaintance and a colleague.) For companies committed to experimentation as a cultural core value not just a useful technical capability, *Uncontrolled* is indispensable reading. I have a particular fondness for Manzi's persistent observation that honest experimentation requires epistemic humility.

Former Bain consultant Richard Koch has repackaged and repurposed his original *The 80/20 Principle: The Secret of Achieving More with Less* (New York: Currency Books/Doubleday, 1998) into a series of popular publications, up to and including a business manga comic book. But repetitions and redundancies notwithstanding, his relentless focus on Vilfredo Pareto's empirical insight that a quintile of causes and inputs typically explain the vast majority of outputs and results is immensely valuable. It transformed how I facilitate hypothesis and experimental design. In this approach, 80/20 alternatives offer economically provocative paths between the normative organizational desire to design optimal experiments (i.e., the best possible) as opposed to just adequate/satisificing experiments (i.e., ones that are good enough).

Abandoning both the optimal and the satisfactory in favor of experiments that seek to identify the vital few causes and essences proved liberating. Koch's explication and extension of Pareto may initially appear simplistic. But his 80/20 framework can instantly discipline wayward design, development, and innovation deployment debates. His books are lightweight, but their impact is heavy.

Quirky (yes, it features comics), comprehensive, opinionated, and personal, Aaron Brown's *Red-Blooded Risk: The Secret History of Wall Street* (New York: Wiley, 2011) appears to be an intimate history of how the quants took over Wall Street. But that characterization obscures Brown's larger and more penetrating contribution: this book really tells the tale of how the tools, techniques, and technologies of risk management came to Wall Street. This is the book to read if you want to grasp the mathematical, statistical, and probabilistic constructs of what measurable risk means to investors. Needless to say, Brown overwhelmingly does this in the context of finance. But the risk management metrics, insights, and heuristics that he explains map beautifully and brilliantly to the investments in experiments discussed in *The Innovator's Hypothesis*.

Because he was a hands-on risk manager himself, Brown understands and fluently communicates the challenges of reconciling risk metrics with risk culture. In virtually every chapter, he explores and explains how probabilistic techniques for identifying, measuring, and mitigating risk co-evolved with the real-world investment frictions and crises. His historical review of the meaning and management of "value at risk" metrics, for example, is masterful. Similarly, his chapter "When Harry Met Kelly," which compares and contrasts the portfolio management—i.e., risk management—approach of Nobel laureate-to-be Harry Markowitz with the "Kelly Criterion" of Bell Labs physicist John Kelly, is an intellectual gem. You read Brown not to design better experiments but to understand how to manage the risks of investing in experimentation better.

Stefan Thomke's *Experimentation Matters: Unlocking the Potential for New Technologies for Innovation* (Boston: Harvard Business Review Press, 2003) deservedly remains a classic. It repays rereading. He arguably overstresses the technological aspects of experimentation, but his focus on rapid experimentation/iteration capabilities bears review. While underplaying the creative importance—and creativity—of hypothesis, Thomke's discussion of how organizations fail to learn from experiments (and thus learn to fail) easily could have been written yesterday—or perhaps next week. The book's emphasis on organizational learning and its case studies of resistance present a different sensibility from my own. Thomke's operational orientation is more top-down than bottom-up; *The Innovator's Hypothesis* celebrates small and agile X-teams in friendly rivalry. Differences in perspective aside, *Experimentation Matters* made a compelling case to the C-suite that experimentation needed to become an organizing principle for strategic innovation initiatives. Thomke's argument still stands.

MIT's Eric von Hippel (Thomke's thesis advisor, by the way) pioneered and publicized the notion of lead users as key innovation drivers. His *Sources of Innovation* (New York: Oxford University Press, 1988) and *Democratizing Innovation* (Cambridge, MA: MIT Press, 2005) have inspired my own 5×5 design sensibilities. Where

Sources immaculately documents the role of lead users, I see *The Innovator's Hypothesis* as a tale of lead experimenters—innovators whose hypotheses and experiments have an excellent chance of being mainstreamed into the organization's value vocabulary. Similarly, *Democratizing Innovation* is implicitly an invitation to "democratize business experimentation." Von Hippel's scholarly discussion of user-led innovation also can be read as a discourse about a new economics of experimentation. The dynamics driving von Hippel's innovation scenarios are kindred spirits to those democratizing experimentation opportunities.

What Is Web 2.0? Design Patterns and Business Models for the Next Generation of Software (Sebastopol, CA: O'Reilly Media, 2009), by Tim O'Reilly, the publisher and entrepreneur who coined the phrase, offers a more populist, webified version of von Hippel's user-driven innovation scholarship. O'Reilly's brilliant description of Web 2.0 services becoming "more valuable the more people use them" is pithy, persuasive, and profound. More important, O'Reilly and his company's authors grok better than almost anyone that the Internet and "Internet of Things" are laboratories whose global potential for disruptive experiment has just begun to be tapped by the Amazons, Googles, Facebooks, and Squares. The Internet is less an ongoing experiment in digital architectures than an architecture that enables ongoing digital experimentation. Experiment is so baked into the culture, mindset, and server tools of digital media innovators that it's practically taken for granted. But as Marshall McLuhan observed, "Whoever discovered water, it wasn't a fish." The fact that so many digital organizations take experimentation for granted may create opportunities for those who don't.

Microsoft partner architect Ron Kohavi's papers and presentations on web-based experimentation are enjoyably useful and usefully enjoyable. They're gems. While emphasizing their Microsoft/Bing origins, Kohavi's discussions of the politics of A/B testing, testing at scale, and experimental design are first-rate and immediately applicable. I've found them wonderful primers and reality checks for 5×5 teams with bolder proposals for website experimentation.

He and his coauthors do a superb job of putting business experiments in technical context. Their work (http://www.exp-platform. com/Pages/default.aspx) deserves a broader managerial and executive audience.

Similarly, the article "A Step-by-Step Guide to Smart Business Experiments," by MIT's Duncan Simester and Northwestern's Eric T. Anderson (*Harvard Business Review*, March 1, 2011, http:// hbr.org/2011/03/a-step-by-step-guide-to-smart-business-experiments/ar/1), offers a more analog counterpart to Kohavi's work. Exceptionally well written and accessible, it artfully distinguishes between analytics versus experimentation in facilitating data-driven insights. As much as I like the article, it describes rather than addresses the natural resistance experiments and experimenters confront. Not unlike Thomke, the Simester-Anderson approach seems less interested in empowerment than getting top management to commit to experimentation. That's not a trivial challenge. The article makes clear that tomorrow's innovation culture wars will be waged between the intuitionists and the datanauts.

The most "pop" approach to simple digital experimentation comes from Dan Siroker and Pete Koomen, the cofounders of Optimizely—arguably the web's leading A/B testing platform. Their work was integral to the remarkable electoral and fundraising successes of the 2008 Obama presidential campaign. Their book, *A/B Testing: The Most Powerful Way to Turn Clicks into Customers* (Hoboken, NJ: John Wiley & Sons, 2013) is filled with anecdote, vignette and accessible case examples. A superior introduction to mixing the technical, organizational and design ingredients for actionable innovation outcomes. But as the title makes clear, the Optimizely bias is more towards test than experiment.

For understanding the technical and quantitative origins of "design of experiment," one can do no better than to read *An Accidental Statistician* (New York: Wiley, 2013), the autobiography of George E. P. Box (1919–2013), a "grand old man" of Anglo-American statistics. As much as a Shewart, Deming, Juran, or Feigenbaum, Box was responsible for making "design of experiment" methodologies industrially accessible. Almost any paper by him is readable, and his autobiography is a delight. He coauthored the classic text *Statistics for Experimenters: An Introduction to Design, Data Analysis, and Model Building* (New York: Wiley, 1978), which is a challenging read, and enjoyed the notoriety of his most-quoted aphorism: "All models are wrong, but some are useful."

Little Bets: How Breakthrough Ideas Emerge from Small Discoveries (New York: Free Press, 2011), by Peter Sims, is a quick, smooth, and anecdote-rich read that describes big ideas, with big impact, that began as little bets. Sims stresses the value of both play and design thinking. His observations are very much in the spirit of both my previous book, *Serious Play: How the World's Best Companies Simulate to Innovate* (Boston: Harvard Business Review Press, 1999), and *The Innovator's Hypothesis*. But *Little Bets* is largely based on ex post facto tales of small bets that paid off. With tongue in cheek, I'd have been happy to change the title of *The Innovator's Hypothesis* to *Little Experiments*.

Eric Ries's *The Lean Startup: How Today's Entrepreneurs Use Continuous Innovation to Create Radically Successful Businesses* (New York, Crown Business, 2011) has become a best-selling business book phenomenon. I've seldom seen and heard entrepreneurs so seriously excited about a start-up text as opposed to a success story. But Ries has done fantastically well reengineering Toyota's "lean principles" into action items for entrepreneurs. With tongue less firmly in cheek, I'd seriously consider renaming this book *Lean Experimentation*. Why? Because along critical design dimensions, that's what this book is primarily about. However, the goal and role of experiment portfolios—and the argument that experiments need not be product- or service-focused to be disruptively innovative—is incompatible with mission-critical aspects of the *Lean Startup* message. That said, when I've run 5×5—or, more accurately, 3×3—workshops at lean start-ups, the participants ate it up!

The last two influential books are by economists. *Discovery—A Memoir* (Bloomington, IN: Author House, 2008) is the delightfully quirky and wide-ranging autobiography of 2002 Nobel laureate economist Vernon Smith. He's an unusual—and unusually brilliant—man and innovator. I found his story particularly influential and inspiring because he was largely responsible for making experimental economics respectable and important. He moved the discipline. Since the eras of Marshall, Keynes, Samuelson, and Friedman, economics had been regarded as more an empirical or a blackboard science. With the notable exception of Harvard's Edward Chamberlain, almost no serious economists did experiments. Vern Smith was one of Chamberlain's students. He ended up building his professional life and reputation forcing the economics profession—along with, yes, behavioral economist Daniel Kahneman—to take simple experiments seriously. He won.

Deirdre McCloskey's *The Rhetoric of Economics* (Madison, WI: University of Wisconsin Press, 1985) is another delightful work. Her book explores how economists persuade the public, the policymakers, and each other. It's filled with examples hilarious, outrageous, dishonest, and unintelligible. But McCloskey acutely understands that passion and promise can be even more persuasive than evidence and analysis. Everybody—especially the quantitatively trained—who reads it tells me that it changes how they change minds. It's an excellent read for everyone who recognizes that sometimes experiments need a little help communicating their real value.

The astute reader will observe that virtually all these books and recommended readings have a heavy first-person or autobiographical component to them. They're rooted as much in their authors' personal and professional experiences as their core competencies and research. That's not an accident. *The Innovator's Hypothesis* describes my own observations and experiences in helping organizations creatively experiment with creative experiments. It's both a personal and professional book. I couldn't help but be inspired—and informed—by the observations and experiences of others. I admire them.

INDEX